Discovering Physics

SI edition

Philip E. Heafford, M.A., B.Sc.
University of Oxford Department of Education

 **Longman**

LONGMAN GROUP LIMITED
LONDON
Associated companies, branches and representatives
throughout the world

First edition © Philip E. Heafford 1964
SI edition © Longman Group Ltd 1970

All rights reserved. No part of this publication
may be reproduced, stored in a retrieval system,
or transmitted in any form or by any means,
electronic, mechanical, photocopying, recording
or otherwise, without the prior permission
of the copyright owner.

First published 1964
SI edition first published 1970
3rd impression 1971

ISBN 0 582 21131 X

Printed in Great Britain by
Butler & Tanner Ltd, Frome and London

The photograph on the cover is Crown Copyright reserved

Introduction to the SI edition

Scientists have for many years used the metric system and metric measures in their work. Now teachers of science in many countries have adopted a rationalized, coherent system of units. The old British units, having no metric connection, will no longer be used in schools.

This edition uses those rationalized and coherent units, abbreviations and symbols recommended by the Association for Science Education for use in schools. These are metric units and follow the Système International d'Unités. They are used internationally and have been accepted by examining boards.

Relationships between units are greatly simplified. The system is coherent because the product or quotient of any two unit quantities in the system is the unit of the resultant quantity. For example, when unit length in the system is divided by unit time the quotient is the unit of velocity in the system.

The unit of force, and therefore of weight, is the newton. Care has been taken not to confuse this unit with that of mass, which is the gramme or kilogramme. Although the unit of length is clearly stated to be the metre, the centimetre and kilometre are used where appropriate. Heat, mechanical, and electrical energy are expressed in joules but reference is made to the kilocalorie for those who still wish to retain this unit for measurement of the calorific value of fuels and food.

Units are not written with a capital initial letter, e.g. ampere, volt and watt, but the symbol for a unit named after a person does have a capital initial letter, e.g. W for watt, Hz for hertz and J for joule. No plural form for the symbol of the unit is used, e.g. 65 kg, and 17 J. The index notation and multiples and sub-multiples of units are also used for simplicity, e.g. a velocity of ms⁻¹, an electric resistance of $7 \times 10^6 \Omega$ or 7 MΩ, and an electric power of 3.5×10^3 W or 3.5 kW.

The celsius or centigrade scale of temperature is used and the kelvin scale is explained.

P. E. HEAFFORD
Oxford

Introduction to the first edition

Discovering Physics is an up-to-date and systematic two-year course in Physics. It is in one volume, and it covers the requirements of the various regional C.S.E. Examination Boards. It may also be used in the middle school by pupils who will take 'O' level. It is important to stress that the whole emphasis and sequence of teaching is in line with the recommendations of the Association for Science Education, while the sequence of teaching the study of the electron and its properties follows, in a simplified and shortened manner, the recommendations of the Committee on the Teaching of Modern Physical Science.

There is ample opportunity for practical work from the start. This is one reason why the study of Light is considered first. The book provides ample opportunity and interest for this approach and the fundamental ideas of Force and Energy are simply and clearly explained before the pupil is introduced to a broader study of Heat Energy and Electrical Energy.

Many traditional experiments and irrelevant facts have now given way to more modern concepts and applications of science. The new look and new approach to the curriculum will be found in this book, which not only describes the internal combustion engine, but also the jet engine, rockets, aerofoils, three phase alternating current supplies, a.c. and d.c. generators and motors, radio communication, transistors, television, radar, and the tape recorder. At the same time, calculations in the text and in the examples are simplified so that the main physical principles are not obscured by the unnecessary manipulation of mathematics.

New methods of presenting some subjects are included, for example, waves are treated as a whole—water waves, sound waves, earthquake waves, light waves, and electromagnetic waves—to show their similarities and their differences. The formation of images by mirrors and lenses is explained by methods of drawing, not by calculation.

The correct use (British Standard) is made of the termination of words—a piece of apparatus (resistor, capacitor), a property (resistance, capacitance), an operation or process (induction, conduction), and the measure of a property (conductivity, resistivity).

The layout and design of the page openings show the text and diagrams to their best advantage, and reflect the new look now being given to physics in secondary schools. This treatment does not mean that the work included misses in any way the requirements of the syllabus. On the contrary, pupils will be stretched as well as stimulated over the two years' work in this book, which will take them well up to, and slightly beyond the C.S.E. standard.

It is hoped that the book will inspire pupils to search for the truth, and to find the real explanations for the numerous wonders of this age of science and technology, and that it will give them an insight into the procedures of scientific method as well as some scientific knowledge. It is, however, the teacher who alone can spread the true spirit of science and show that it is not the mere acquisition of facts.

The author is most grateful to all those teachers who have given help and advice, and particularly to Mr M. L. Haselgrove for his careful reading and many helpful suggestions.

PHILIP E. HEAFFORD

Contents

Electrical Energy

List of units, abbreviations, and symbols

absolute zero	0 K	intensity of illumination	E
acceleration	a	joule	J
alternating current	a.c.	kilogramme	kg
ampere	A	kilometre	km
amplitude modulation	AM	kilowatt hour	kWh
angle of incidence	i	length	l
angle of reflection	r	long wave	LW
angle of refraction	r	luminous flux	Φ
audio frequency	AF	luminous intensity	I
brake horse-power	b.h.p.	mass of a body	m
candela	cd	mechanical advantage	M.A.
centimetre	cm	medium wave	MW
centre of gravity	c.g.	metre	m
charge, quantity of electricity	Q	microfarad	μF
coulomb	C	microwaves	UHF
cubic centimetre	cm^3	minute	min
cubic metre	m^3	nanometre	nm
current of electricity	I	negative electric charge	—
distance	s	newton	N
direct current	d.c.	north pole	N.
electromotive force, e.m.f.	E	ohm	Ω
farad	F	picofarad	pF
force	F	positive electric charge	$+$
frequency	f	potential difference	V
frequency modulation	FM	quantity of heat	Q
gramme	g	radio frequency	RF
hertz	Hz	resistance, external	R
horse-power	h.p.	resistance, internal	r
hour	h	resistivity	ρ
index of refraction	μ	second of time	s

short wave	SW	time	t
south pole	S.	ultra high frequency	UHF
specific heat capacity	c	velocity of wave	V
specific latent heat	l	velocity, final	v
square centimetre	cm²	velocity, starting	u
square metre	m²	velocity ratio	V.R.
temperature	T	very high frequency	VHF
temperature change	$T_1 - T_2$	volt	V
temperature, celsius	°C	watt	W
temperature, Kelvin	K	wavelength	λ

Notes 1. Symbols for units do not have a plural form with an added 's'.

2. Numerical quantities are expressed in powers of 10—either positive or negative powers.

3. 'per' is not used with symbols, e.g. speed is written in units of m s⁻¹.

4. When abbreviations like a.c., e.m.f. begin a sentence, capitals are used throughout, e.g. A.C., E.M.F.

1. The nature of light

Some objects are capable of changing various forms of energy into light energy. The sun is one of these objects. They are said to be *luminous*. Other objects, such as wood, iron, and flowers are *non-luminous*; they simply reflect the light energy that falls upon them from some other source and are then said to be *illuminated*.

When solids and liquids are heated to temperatures greater than 800°C they give out light and then they are said to be *incandescent*. For instance, if you look closely at the light sent out from a fire or a candle you will see that it really comes from a large number of small hot particles in the flames. These particles cease to give off light when they cool down at the edges of the flame or in the air above it. In fact, the brightness and colour of the natural light from these incandescent particles varies with their temperature.

However, light can be given off by many different kinds of *fluorescent* lamps without the sources becoming incandescent. It is interesting to note that the colour of the light given off by such a lamp does not depend on its size or brightness. In such artificial sources the light is not due to hot particles but is in some way derived from the electrical energy supplied to them.

An object can only be seen by the eye because of the light that travels from the object to the eye. An object may either emit the light itself or it may reflect some of the light that illuminates it coming from another source.

When light travels through some material it emerges at the other side in different ways according to the quality of the material. If the material is clear sheet glass then the light is hardly disturbed and the glass is said to be

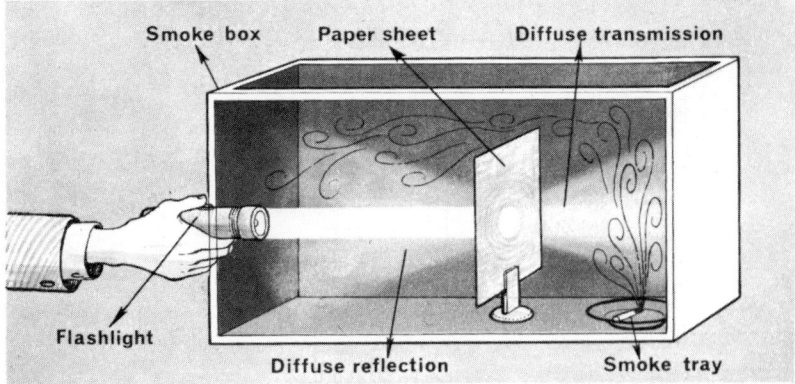

Figure 1.1. A smoke box in use. The small quantity of smoke given off by the smouldering cigarette in the tray enables us to see the direct beam of light and the light diffused by the paper.

1

transparent. If the material is coloured glass it is transparent to that colour but *opaque* to the other colours of the spectrum. Materials like this sheet of paper *diffuse* the light and cause it to scatter on both sides of the paper. To be more precise if this piece of paper were held in front of a bright light there would be *diffuse reflection* of the light on one side of the paper and *diffuse transmission* through to the other side.

The shadow formed when light from a point source strikes an object is sharp and it outlines accurately the shape of the object. The reason for this observation is that light travels in straight lines.

Light given off from a source of reasonable size compared with an object forms a shadow that is blurred at the edges. The region where the shadow is blurred is called the *penumbra*, and is the area where the illumination on the screen gradually changes from full brightness to complete darkness. The region in the shadow of complete darkness is called the *umbra*.

During an eclipse of the moon it is possible to follow the path of the moon as it passes from full brightness through the earth's penumbra until it is fully eclipsed in the umbra. Sometimes, however, the moon is not sufficiently in the direct line of the sun and the earth for the whole of it to pass into the umbra of the earth's shadow. When this happens only a *partial eclipse* of the moon is seen from the earth. For a *total eclipse* to occur the moon must pass entirely through the umbra.

The shadow of the moon falls upon the earth when an eclipse of the sun takes place. A person may experience during that time what it is like to be in the penumbra of the moon, and if he is fortunate, what it is like in the region of the umbra also. He notices as the eclipse progresses how the light gradually diminishes in the region of the penumbra and then how it becomes very dark in the umbra. As he looks at the sun through smoked glasses he sees the sharply defined edge of the moon slowly biting its way across the surface of the sun. That is why the illumination of the earth around him diminishes. Finally when the umbra reaches his position all the direct rays of sunshine are stopped by the

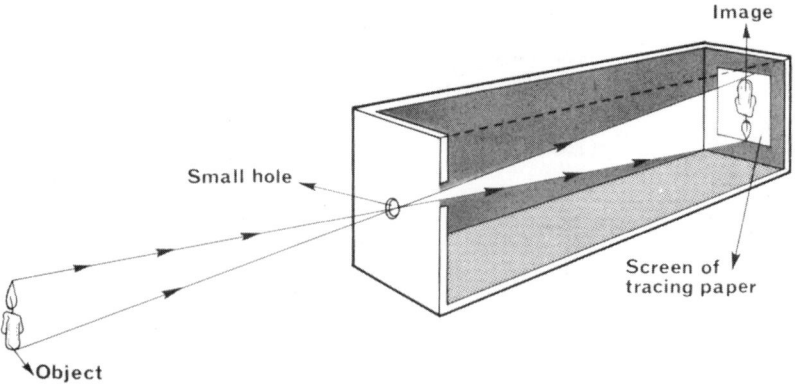

Figure 1.2. A 'pinhole' camera. Light travels in straight lines from points on the object to corresponding points on the image. A perfect image is formed if the pinhole is a small one.

Figure 1.3. This boy is producing 'dog shadows' with his hands and fingers. The source of light on the left is small and strong. The size of the shadow formed varies according to the relative positions of the screen, the boy, and the source of light.

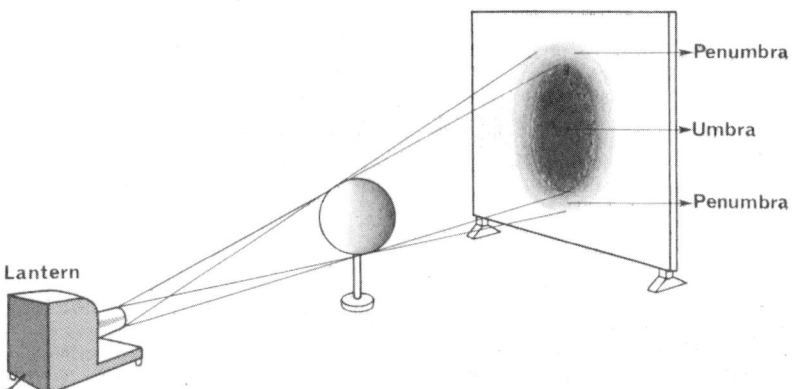

Figure 1.4. This shows how light from a large source produces a shadow that is ill-defined. The larger the source the wider the area of the penumbra.

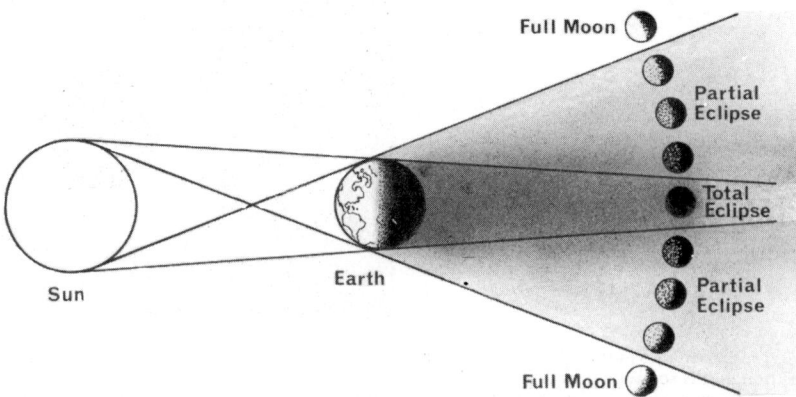

Figure 1.5. The various stages during the total eclipse of the moon. The scale in the diagram has been falsified to make these stages appear more obvious.

moon and he then gets a magnificent view of the sun's corona. These great flames shooting out from the surface of the sun, called the corona, are only visible when the sun's surface itself is completely hidden by the moon. The watcher, if he observes carefully, will note also that the heat from the sun ceases during and recommences after the eclipse at the same moment as the light. Heat therefore travels at the same speed as light over these enormous distances. This is an important observation that we shall refer to later in this chapter.

The detection of light is made possible by using one of several instruments —the most common is *the eye*, if the eye can be called an instrument. The eye is extremely adaptable and sensitive to the rays of light. Have you ever noticed how sharply things stand out on a bright moonlit night? Yet, in the full strength of midday sunshine in summer the same eye responds easily and sees distant objects clearly even though the illumination may be more than a hundred million times as bright as it is in moonlight.

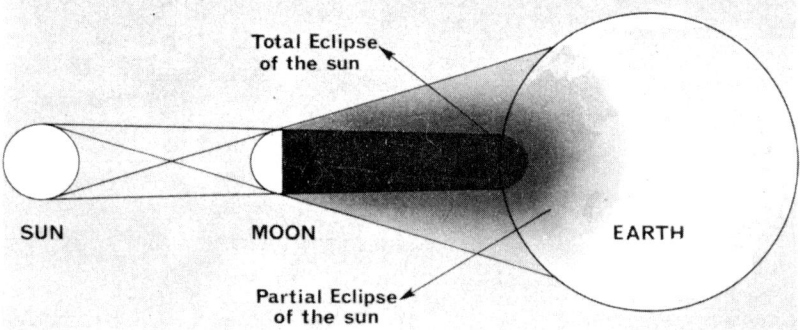

Figure 1.6. An eclipse of the sun showing the positions on the earth from which the total eclipse and the partial eclipse are visible. The whole shadow pattern moves quickly over the earth's surface in one long arc so that it is visible only by those who are in its track. This diagram is not drawn to scale. The umbra normally covers a smaller area than it appears to do here.

Figure 1.7. Stages in the eclipse of the sun as seen from one position on the earth. The moon, although it is so much nearer the earth, appears to be only slightly bigger than the sun and only just manages to cover the sun's surface at total eclipse. When a total eclipse occurs it lasts only a few minutes during which time the corona can be seen clearly.

Another instrument for detecting light is a *photo-electric cell*, which consists essentially of a plate covered with a substance that gives off electrons when light falls on it and it is connected in an electrical circuit. Many substances have this property, some more than others—they include sodium, potassium, lithium, selenium, calcium, and copper oxide. The stream of electrons flowing from a photo-electric cell depends on the strength of the light falling on the cell. Practical use is made of such cells in the sound reproduction system of the moving film, and in the working of automatic light switches.

A *photo-voltaic cell* made by coating an iron plate with selenium produces a very small electric voltage when light falls on it. A *light meter* consists of a photo-voltaic cell and a sensitive galvanometer by means of which the voltage generated is measured.

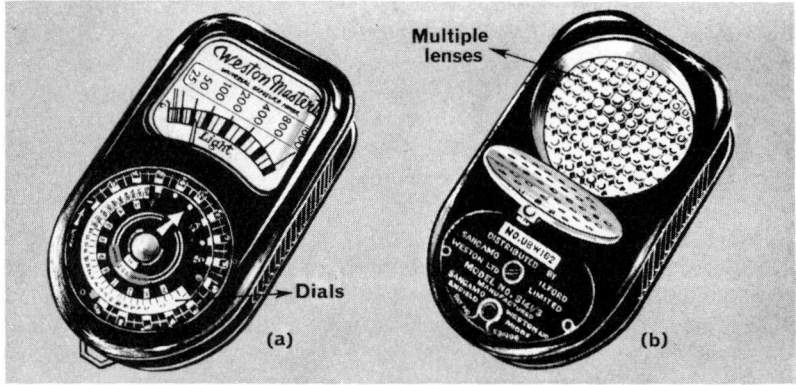

Figure 1.8. A photographic exposure meter. (a) The pointer indicates the intensity of illumination falling on the lenses on the reverse side of the meter. The circular calculating dials enable this light value to be converted easily into the correct exposure time. (b) The honeycomb type of multiple lenses beneath which lies the light sensitive photo-voltaic cell of the meter.

Some salts of silver are *photosensitive* and are used on the surface of photographic films and papers. These silver salts decompose when struck by rays of light and after chemical 'developing and fixing' form finely divided particles of metallic silver. These are the dull black portions that appear where the light fell on the film or paper in the making of the negative or the positive in the *photographic process*.

Some photosensitive chemicals and photo-electric cells can detect rays that cannot be seen by the eye. We know that these are rays similar to light rays because, like light rays, they can be reflected and refracted. We know too that they travel with the light rays from the sun and are cut off when the light rays are cut off. Their presence can be detected also because they bleach some materials, they turn the human skin brown, they kill bacteria, and they enrich the colour of the vivid paints often used in outdoor advertisements. These are called *ultra-violet rays.*

There are other rays like light rays that can be detected by their action on certain chemicals and by photo-electric cells. They warm objects placed in their path and are called *infra-red rays.* Our skin can feel the rays and yet our eyes do not see them. If you switch on an electric radiant fire and put your hand in front of it you will feel the heat of the infra-red rays long before you will see the wires becoming red or even dull red. Infra-red rays can pass through some materials that are opaque to light rays. For example, we can with infra-red rays take photographs of distant scenes that cannot be seen with light rays because infra-red rays are not scattered as light rays are by haze and dust in the atmosphere. Photographs can even be taken by infra-red rays in a totally dark room. Photo-electric cells can be made that are very sensitive to infra-red rays. These cells enable us to build counting devices, automatic selectors, door openers, and burglar alarms.

The speed of light. Rays of light travel rapidly as we observe when a flash of lightning takes place or when someone signals from a distant hill—how rapidly we can only determine by timing the passage over long distances. It is remarkable that as long ago as 1676 the Danish astronomer Roemer, observing the satellites of Jupiter, was able to calculate a value for the speed of light that for all normal purposes we use today, namely 3×10^8 m s^{-1}. This is the same speed that we have determined for the passage of an electric

Figure 1.9. A black and white photographic process. (a) A film negative after developing and fixing—the sky is black and the people are light. (b) A paper positive made by passing light through the film negative and developing and fixing the exposed photographic paper.

current along a wire and for radio waves. Radio waves, infra-red rays, visible light, and ultra-violet rays all travel at the same speed and through space without the help of or being carried by any material in that space.

What do we know of *the nature of light*? It is a form of energy because some other form of energy must be used up to produce a source of light. We shall learn later that if an amplifier is connected to a photo-electric cell on which light is falling an irregular clicking sound is heard. This suggests that something is being ejected by the cell in particles one by one. One theory is that light is composed of small particles of energy each called a *photon* or a *quantum*, and that when these photons score a 'hit' an electron is knocked out of the cell. This is the particle or *corpuscular theory*. Another theory is that light has the nature of a wave because it travels and is reflected and refracted as waves. This is known as the *wave theory* of light. The different colours of light can be explained on this theory by considering that each colour has a different wavelength. If we accept this then we can give to the infra-red rays a longer wavelength than red light and to ultra-violet rays a shorter wavelength than violet light.

These two theories of light—the corpuscular theory and the wave theory present scientists with a challenge. Scientists now think that the true explanation may be a combination of the two. We can accept for the time being the limitations and explanations that both theories offer and apply them to the understanding of our observations of the nature of light. Like all true scientists we must be prepared to abandon these theories as soon as a better and more comprehensive explanation is forthcoming.

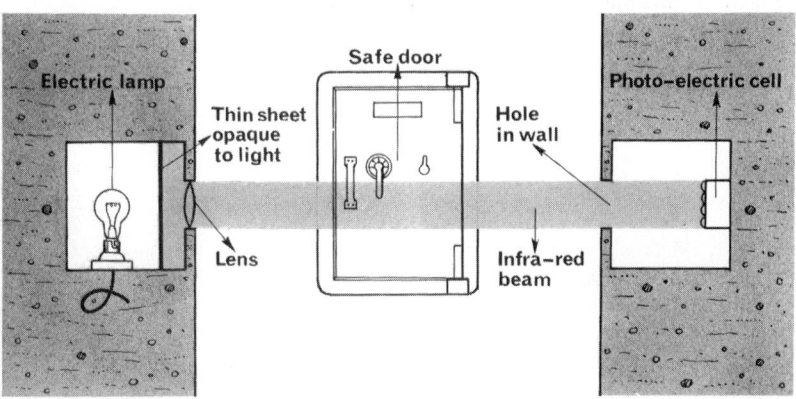

Figure 1.10. A burglar alarm installed in front of a safe. The invisible beam of infra-red rays if interrupted by a burglar approaching the safe would cause the photo-electric cell to operate an alarm bell. The thin sheet in front of the electric lamp in the projector is opaque to light rays but transparent to infra-red rays.

B

Questions on chapter 1

1. Draw a diagram to show how an image is formed in a pinhole camera. Describe the image. What is the effect on (a) the clarity of the image, and (b) the length of the exposure needed, if the hole is enlarged?

2. A penny held with one of its flat sides facing a large electric lamp is gradually moved further from the lamp and nearer to the wall of a room. What changes will take place in the shadow cast by the penny on the wall? Give reasons for your answer and draw two or three diagrams to illustrate what you mean.

3. Under what conditions is the total eclipse of the sun possible? Why is any particular eclipse of the sun visible only over a small area of the earth's surface?

4. Why is it that the corona of the sun can only be seen by the eye, looking through a smoke glass to protect it, during the periods of the total eclipse of the sun?

5. A dark room whose floor is 4 m square has a hole in one of its white paper-covered walls. Outside the wall with the hole in it and at a distance of 30 m is a lighted Christmas tree. The image of the tree appears on the wall of the room opposite the hole. How many times taller is the lighted Christmas tree than its image?

6. The moon has a diameter of 3 584 km and is 384 000 km away from the earth. How far from the eye must you hold a disc of 1 cm diameter so that the moon is just obscured by it?

7. A man is 2 m tall and stands 4 m away from the base of a wall. On the wall a lamp 4 m high above the ground is shining. How far is the shadow of the head of this man from the base of the wall?

2. Reflection

Plane mirrors

Plane surfaces such as metal or glass mirrors, the smooth surface of a lake or river reflect light in regular symmetrical ways. It can easily be shown that the normal (a line perpendicular to the surface) and the two rays of light are always in the same plane surface. It is only necessary to observe these rays as you direct a beam of light from a flash-lamp on to a mirror and then move the beam about—up, down, and sideways. Dust or smoke in the air will make the track of the rays visible. A smoke-box will enable you to do this experiment (see Figure 1.1). If you keep the flash-lamp fixed you can then measure the two angles formed between the incident ray and the normal and between the reflected ray and the normal. Repeat this in several different positions. Do these experiments enable you to state any laws (conclusions that are always true) about the incident and reflected rays of light when they strike a plane mirror?

The image formed by an object in a plane mirror can be seen but not touched. You will not burn your finger by placing it above the image of a burning candle. Such an image is called a *virtual* image. It can be proved by mathematics that the virtual image is just as far behind the mirror as the object is in front. One always measures these distances perpendicularly to the mirror.

Look at yourself in a plane mirror. Raise your right hand to your right ear. You will notice that your image appears to touch its left ear with its left hand. Look at some printing reflected in the same mirror—the image of the printing is reversed. This is called *lateral inversion*. Does inversion take place vertically at

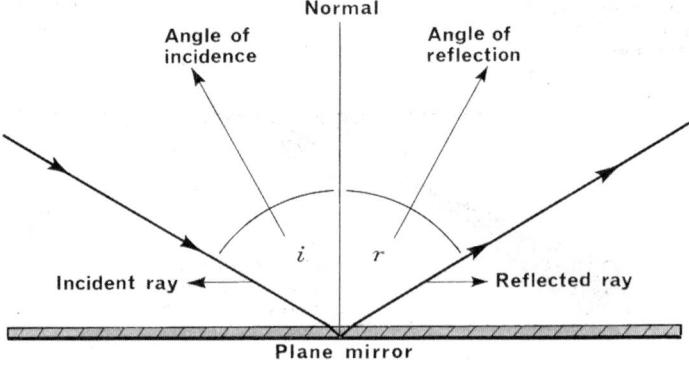

Figure 2.1. The laws of reflection state that (1) the angle of incidence equals the angle of reflection ($\angle i = \angle r$), and (2) the incident ray, the normal, and the reflected ray lie on the same plane surface.

9

the same time? How can you arrange the mirror to give vertical inversion, i.e. so that the top of an object appears to be at the bottom?

Some instruments form their images by reflecting the rays of light in one plane mirror first and then afterwards in another plane mirror. We shall now consider some of these instruments.

The *periscope* has two mirrors parallel to one another and placed some distance apart. On a large sheet of paper draw lines to represent the two mirrors of a periscope. Now draw the path of two rays, one in red ink and the other in blue ink, that travel from a distant object, are then reflected in the two mirrors, and finally enter the observer's eye. How many times does lateral inversion occur? Is the image formed in the periscope inverted or upright?

Two mirrors at right-angles produce three images of an object placed in between them. It is interesting to note the symmetry of the design so formed— the object or an image in each quadrant.

The *kaleidoscope* has two plane mirrors set at 60° to one another and mounted in a tube. In this case a six-fold symmetrical pattern is formed of any object placed between the two mirrors. The images are formed by multiple reflections in the two mirrors in the same way that images are formed in two mirrors at right-angles.

At what angle must the mirrors be set to give an eight-fold pattern? Calculate the answer to this before you observe it by experiment.

Designers use kaleidoscopes when they search for pleasing symmetrical patterns.

Two *parallel plane mirrors*, sometimes seen in a shop or restaurant, produce an infinite number of images. As the number of reflections increases so the illumination of the image gets weaker and the distance the image is away from the object gets greater. Eventually it becomes difficult to see an image at all.

When rays of light strike *irregular surfaces* and not plane ones, as we have considered so far, no regular patterns or images are formed. The rays are reflected in every direction overlapping and intermingling with one another.

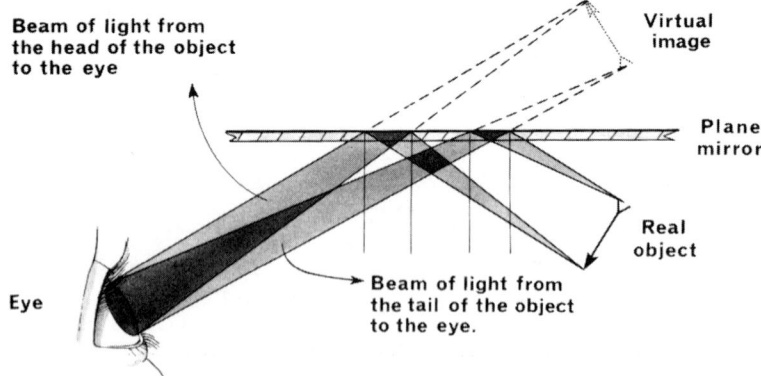

Figure 2.2. This diagram shows how an eye sees a virtual image, which is laterally inverted, of an object in a plane mirror. Only the two extreme points are considered in this diagram— others would confuse the construction lines. The faint lines are the normals. Note that about any one of the normals the angle of incidence is equal to its angle of reflection.

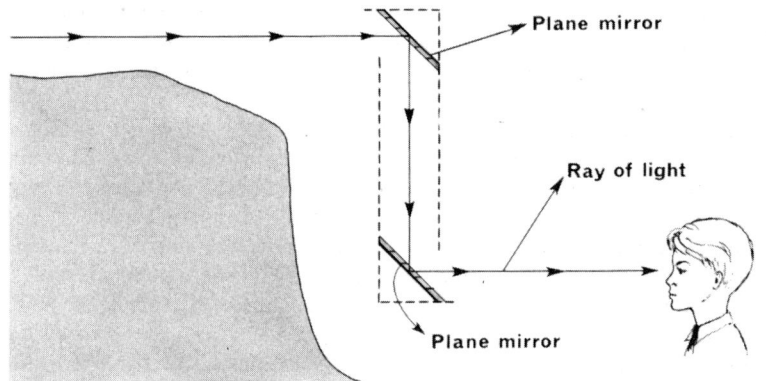

Figure 2.3. How the periscope works. Objects that are hidden from view by an obstruction can be seen by the observer. Can this instrument be used horizontally as well as vertically?

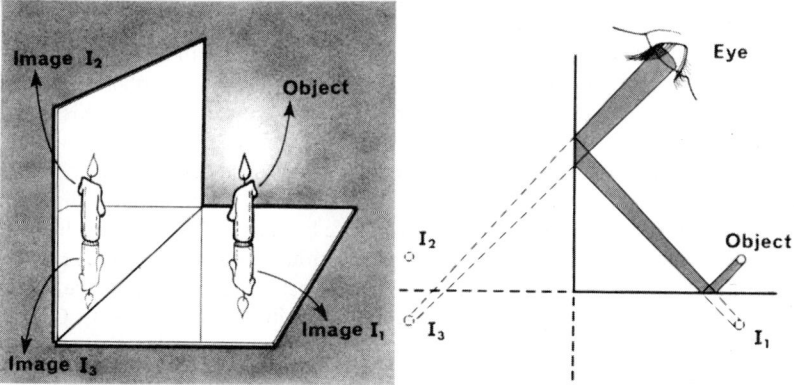

Figure 2.4. A candle stands on one of two mirrors at right-angles. All three images are plainly visible from our viewpoint. Images I_1 and I_2 are each formed by the normal reflection of the light rays in one or other of the mirrors. Image I_3 is formed by reflection first in one of the mirrors and then in the other mirror as shown in the diagram. Does it matter which mirror is the first to reflect the light rays to form the image I_3?

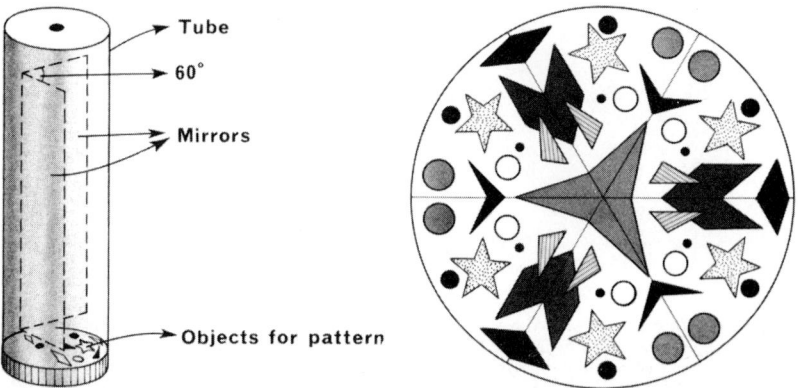

Figure 2.5. How to construct a simple kaleidoscope, and an example of a six-fold symmetrical pattern seen through it.

For instance, the rays of light that strike this page of the book do not form a clear image of the lamp that is their source of light. The paper appears white because most of the light that strikes the paper is reflected, although the reflection is *scattered*.

Scattered reflections also occur when light rays fall on minute irregular particles. If, for example, a beam of light is projected through a dust-laden atmosphere the path of the beam can be seen clearly because of these scattered reflections. But if the atmosphere is absolutely free of particles the beam cannot be seen from a position at right-angles to its direction of propagation. When the particles that cause the scattering are extremely small the scattered light is coloured blue. That is why dry smoke and chalk dust appear to be blue in sunlight. At the setting of the sun the light reflected from the particles in the atmosphere causes twilight and gives the blue colour to the sky overhead.

Curved mirrors

A *concave mirror* has its reflecting surface on the inside of the curve facing towards the *centre of curvature*. For the purpose of the diagrams that follow large sections of concave mirrors that are part of spheres will be used. In many practical applications the mirrors are part of paraboloids, not spheres, but the theory is almost the same, especially when the rays pass close to the axis of the mirror.

On pages 14 and 15 we have a diagram for every possible position of an object placed in front of a concave mirror. If you examine carefully each diagram you will see that two rays from the object have been used. The image is formed at the place where these two rays cross each other after reflection. There are four different rays that can be used for the construction of the image of an object, but the actual rays travel along an indefinite number of paths. One of the rays used for the construction of the image travels parallel to the *principal axis* and is reflected through the *focus* (F), one passes through the focus and is reflected parallel to the principal axis, one passes through the centre of

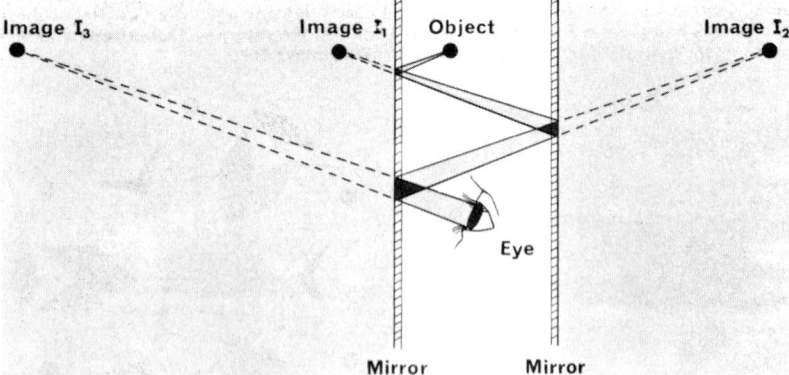

Figure 2.6. How an eye sees the third image of an object in two parallel plane mirrors. Which images are laterally inverted and which are the right way round? Can a person sitting between two parallel mirrors see the back of his head?

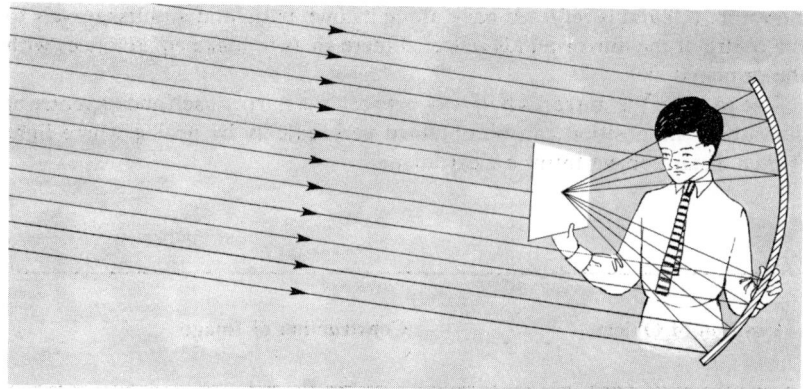

Figure 2.7. This boy is finding the focal length of a concave mirror by focusing the light from the sun on to a sheet of white paper. The focus is at the place where the clearest image is formed, and the focal length is the distance from the focus to the mirror. The heat rays from the sun also converge to the same focus and may cause the paper to burn.

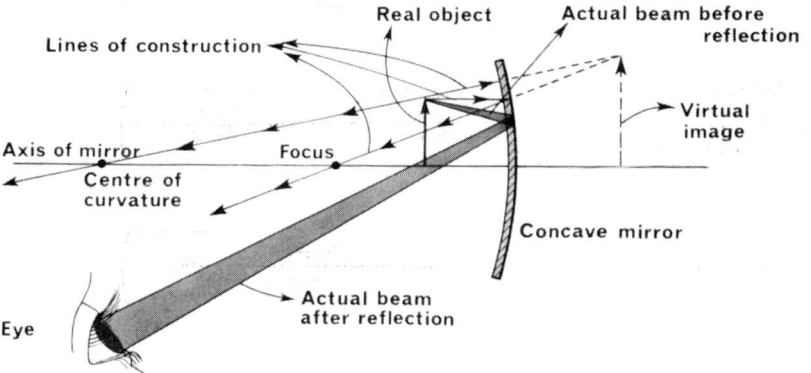

Figure 2.8. This shows how the beam of light from a point on an object travels to the eye after reflection in a concave mirror. The eye sees an enlarged, erect, and virtual image of the object.

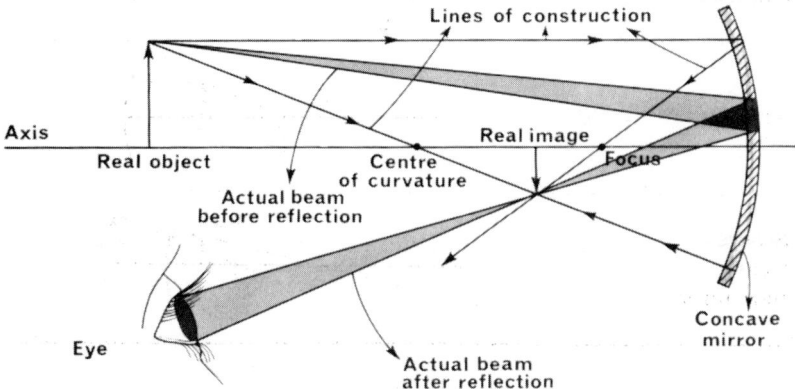

Figure 2.9. This shows how an eye sees a smaller, inverted, and real image of an object after reflection in a concave mirror.

curvature (C) and is reflected back along its own path, and another travels to the centre of the mirror and is reflected there so as to make equal angles with the principal axis.

The focus of the mirror is half way between the mirror itself and its centre of curvature—its position can be obtained very quickly by noting where light from a distant object forms a clear image.

Position of Object	Construction of Image
At infinity	
Between infinity and centre	
At centre	
Between centre and focus	
At focus	
Between focus and mirror	

These diagrams also tell us the position of the images, their relative size, and if they are real or virtual images. There is no need to remember each of the diagrams because it is an easy matter to reconstruct any one of them by following carefully the path of the two chosen rays.

Now let us draw an actual beam of light as its starts from the object, is then reflected in the mirror, and eventually enters the eye. The eye in every case will see the object in the image position and will consider it to be there. Two cases will be drawn. Try to do the others for yourself.

Image			Practical Uses
Position	**Relative Size**	**Nature**	
At focal point	Smaller	Inverted Real	Astronomical reflecting telescope.
Between focus and centre	Smaller	Inverted Real	Terrestrial reflecting telescope.
At centre	Same	Inverted Real	Mirror at the back of lamp in a projector.
Between centre and infinity	Larger	Inverted Real	Floodlight. Projection type of television.
Infinity		No image— parallel beam	Searchlight. Car headlight. Pocket torch.
Behind mirror	Larger	Erect Virtual	Floodlight. Shaving mirror. Dentist's mirror for use in mouth.

One of the clever things that has been designed in recent years is the *dipping headlight* fitted to motor-cars. The lamp itself has two filaments—one that is lighted for the forward beam and the other for the dipped beam. One filament is at the focus of the mirror and the other is deliberately offset to cause the major portion of the beam to be tilted down and at the same time to the side of the road.

A *convex mirror* has its reflecting surface on the outside of the curve. The centre of curvature is behind the mirror and the focus is, as in the case of the concave mirror, almost exactly half way between the centre of curvature and the mirror itself.

There is only one construction needed to find the image of an object reflected in a convex mirror. The size of the image varies with the position of the object but the image is always erect, smaller, and virtual. This type of mirror is used on a motor-car to enable the driver to observe a wide angle of the road behind him.

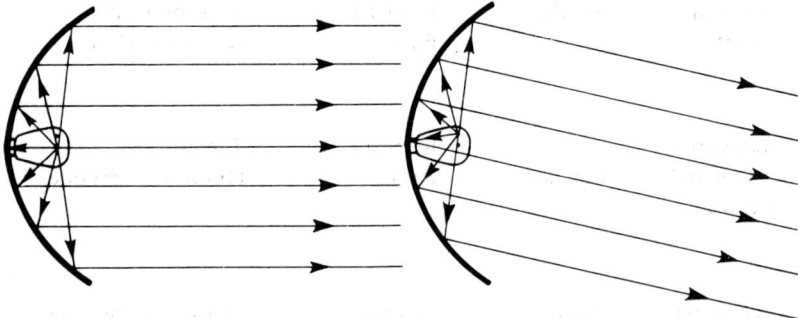

Figure 2.10. A two-beam headlight for a motor-car. These diagrams show how the beam can be tilted down. A similar offset position causes the beam to be deflected to the left if we travel on the left-hand side of the road.

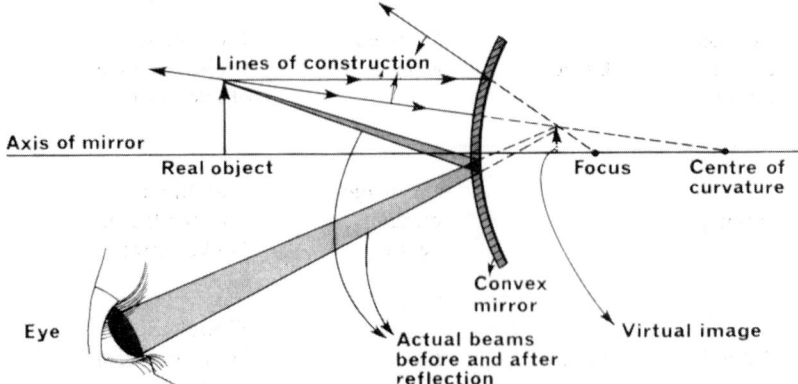

Figure 2.11. This shows the path taken by a beam of light from a point on an object, placed in front of a convex mirror, to an eye. The image is smaller than the object, is erect and virtual. Notice that the rays of light drawn are the same as for the concave mirror—one through the centre of curvature, the other parallel to the axis and then made to pass through the focus. Since light cannot go *behind* the mirror the sections of these rays behind the mirror are drawn dotted.

Questions on chapter 2

1. State the laws of reflection of light. Describe how you would find the image of a single object in a plane mirror. What do you know about the position of this image? Illustrate your answers with diagrams.

2. A plane mirror is set up vertically on the wall of a room. A person stands 1 m in front of it and looks towards the mirror to see the image of the opposite wall of the room. What is the minimum length of the mirror for the person to be able to see the complete height of the opposite wall if it is 4 m high and the distance between the walls is 7 m?

3. A horizontal beam of light falls on a vertical plane mirror which revolves about a vertical axis in its own plane. Show by diagrams that the reflected beam revolves at twice the rate of the mirror.

4. Find, by scale drawing, the size and position of the image of an object 4 cm high placed 25 cm in front of a concave spherical mirror of focal length 10 cm.

5. Find a simple equation that will enable you to calculate the ratio of the size of the image to the size of the object when reflection takes place in a concave mirror. You are provided with the distances of the object and the image from the surface of the reflecting mirror.

6. Draw a diagram to illustrate one of the ways a motorist can dip the beams of light emitted from the headlamps of his car.

3. Refraction

Light may be bent or *refracted* when it passes from one material or *medium* to another. We experience this on many occasions in our daily lives. Think of the apparent bending of a rod or stick that is standing at an angle from the vertical in a pond of water. Look at some printing on a piece of paper through a block of glass and then slowly turn the glass. A most vivid experience is to float slowly along in a small boat over a shallow pond and to watch a weed near the bottom of the pond appear to fall as you go over it and then rise as you go away. You can see the same thing as you walk along the edge of a swimming pool and look at a guide line marked along the bottom of the pool. People are often misled into thinking that the water is not as deep as it really is, thus the notices 'Deep Water' are erected to warn the non-swimmers.

When a light ray passes at an angle from a rarer medium, such as air, into a denser medium, such as glass, the ray bends towards the normal drawn at the point where it enters the glass. This is refraction taking place, and the amount of bending varies with the angle of incidence and the kind of materials that compose the rarer and denser media.

A ray of light on refraction has been observed to obey a certain law. This is known as the *law of refraction*, or sometimes *Snell's law*, because it was Snell who first discovered in 1621 a relationship between the angle of incidence and the angle of refraction.

The ratio of the lengths of the perpendiculars $\dfrac{AN_1}{BN_2}$ and $\dfrac{CN_3}{DN_4}$, using the

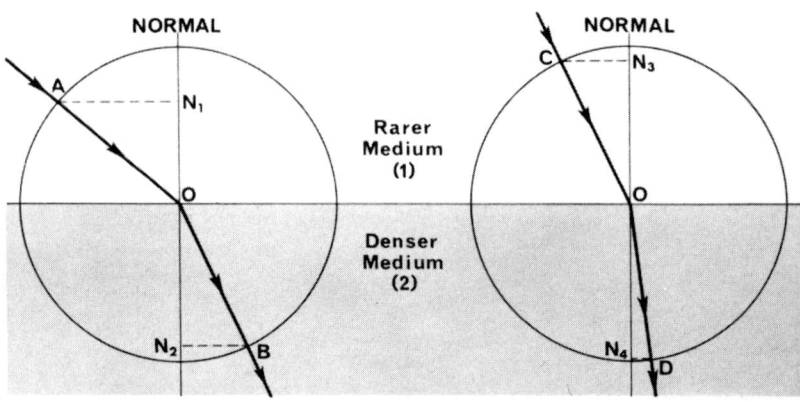

Figure 3.1. Refracted rays of light. In these diagrams two rays are shown incident at different angles as they pass through the surface separating one medium from another medium. Circles are drawn with their centres at the points where the rays cut the surface between the two media. Perpendiculars are then drawn to the normals.

letters as in Figure 3.1, is a constant for any incident ray of light. This relationship is expressing Snell's law of refraction. Another way of expressing it is by using the trigonometrical ratio of the sine of an angle.

$$\frac{\text{sine of the angle of incidence}}{\text{sine of the angle of refraction}} = {}_1\mu_2$$

The constant ${}_1\mu_2$ is known as the *index of refraction* for light passing from medium 1 to medium 2. The index of refraction from air to glass is about 3/2 and from air to water is about 4/3.

Now let us examine a light ray passing in the reverse direction from a denser medium to a rarer medium.

Let us consider the light rays reflected from a weed below the surface in a still pond and travelling upwards to the water-air surface. Figure 3.3 shows four such rays (2), (3), (4), and (5). The angles of incidence of all these four rays are smaller than their respective angles of refraction. The reason for this is that these rays start in a denser medium and emerge into a rarer medium. The rays illustrated in Figure 3.1 both travelled from a rarer to a denser medium.

The ray marked (1) in Figure 3.3 does not suffer any refraction and travels in a straight line through the water-air boundary. The angles of incidence and refraction in this case are both zero.

The ray marked (5) skims along the surface when it emerges into the rarer medium and is called the *critical ray*. The angle of incidence in the denser medium between the ray itself and the normal is known as the *critical angle* (*c*). Any other ray, such as that marked (6), whose angle of incidence is greater than that of the critical ray is *totally internally reflected*. Whether a ray is refracted or totally internally reflected depends on whether its angle of incidence is smaller or greater than the critical angle.

The last diagram in Figure 3.2 shows how the critical ray skimming along the surface boundary enters a denser medium from a rarer medium.

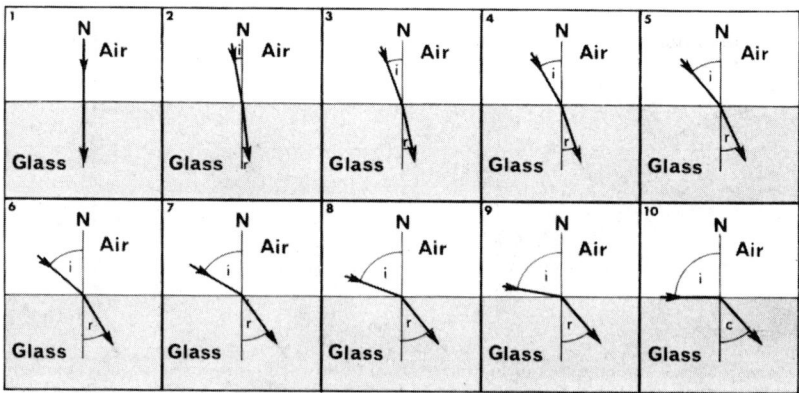

Figure 3.2. Ten different light rays are shown in these diagrams entering the same denser medium at increasing angles of incidence (*i*). Note that the angles of refraction (*r*) increase steadily until in the last diagram the angle of incidence becomes a right-angle and then the angle of refraction is called the critical angle (*c*).

The reason why a weed, fixed firmly by its roots to the bottom of a pond, appears to sink lower as you glide over it in a boat is that the light rays are refracted, as described above, and appear to have come from different positions in the water in the pond.

You must have noticed that a straight stick or rod appears to be bent when it is partly immersed at an angle to the normal in a container of water. The bending takes place at the surface of the water. The explanation of this observation is the same as that for finding the position of the image of the weed in a pond. The refraction of the light from the stick causes the observer to imagine that the stick is higher than it really is.

Total internal reflection. What has happened to ray number (6) in Figure 3.3? The angle of incidence in this case is greater than the critical angle (marked c in Figure 3.3) which is $48\frac{1}{2}°$ for light passing from water to air. When this happens the incident ray is completely reflected. This reflection inside the water is called *total internal reflection*.

We should note that the same principles of refraction and of total internal reflection occur at air-glass surfaces as at air-water surfaces. The only difference is, as we noted before, that the amount of bending is different and that the critical angle is different.

Total internal reflection in right-angled prisms has many interesting and useful applications. The critical angle for light passing from glass to air is $42°$ so that any ray that strikes a glass-air surface at an incident angle greater than $42°$, and this includes any ray that strikes it at $45°$, is totally reflected. If you look at the right-angled reflecting prism drawn in Figure 3.6 you will see that the incident rays pass straight through one side (i.e. along the normal) and strike the hypotenuse at $45°$. Therefore they are all reflected and pass out of the prism

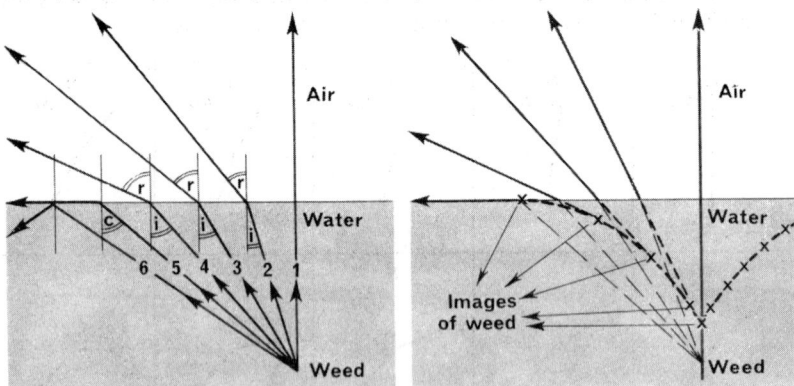

Figure 3.3. Six different rays of light are shown in this diagram leaving a weed below the water surface of a pond. Five are refracted into the air although one (1) emerges straight through the surface boundary undisturbed and another (5) skims along the surface. The sixth ray (6) is reflected back into the water itself and never enters the air at all.

Figure 3.4. This diagram shows how the image of a weed in a pond descends as it is seen by a person moving over it in the air above. The image moves along a line called a 'caustic' and is deepest when the person is immediately above the weed. The actual paths of the rays in air are shown by the continuous lines, and the dotted lines are the paths along which these rays appear to have come in the water.

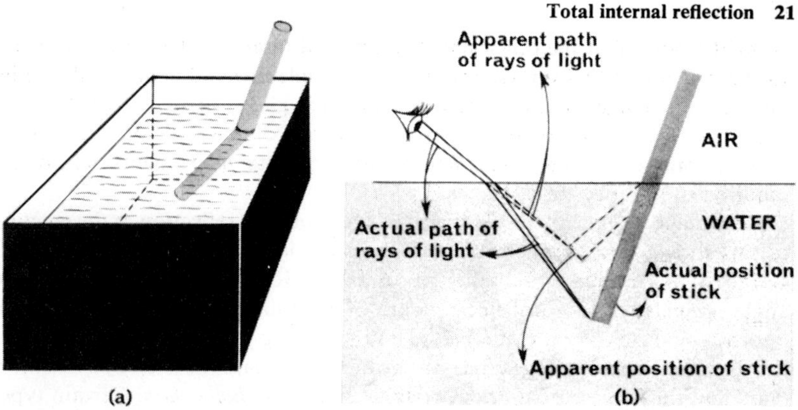

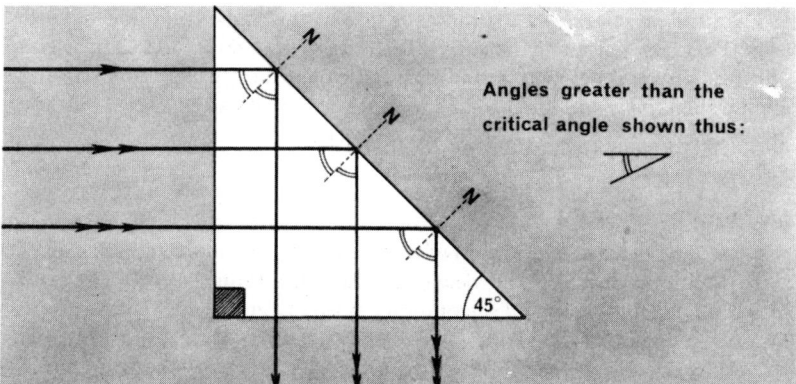

Figure 3.5. (a) The appearance of a straight stick when it is partly immersed in water (b) A cross section showing the paths of the light rays from a point at the end of the stick.

Angles greater than the critical angle shown thus:

Figure 3.6. A right-angled isosceles glass prism (angles 45°, 45°, and 90°) forms a perfect reflector of rays of light.

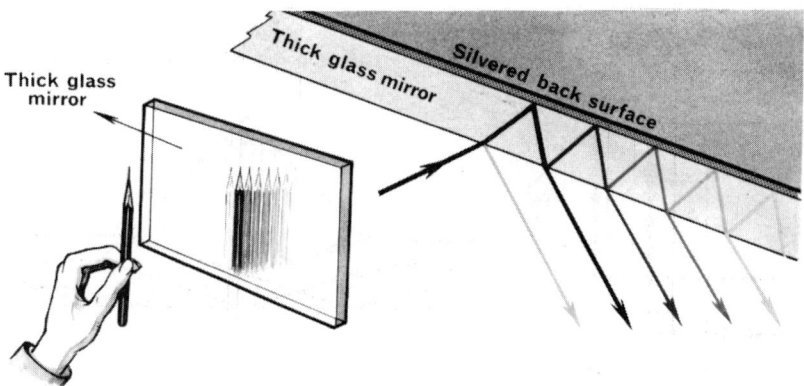

Figure 3.7. Thick glass mirrors form multiple images of objects seen in them due to refractions and to reflections from the top and bottom surfaces of the mirror. Why is the second image the strongest? Why do the images become widely separated when the object is reflected and seen from the side?

along the normal at the other side. Perfect total internal reflection takes place so that the image this surface produces is a single one and is not confused by the multiple images that are produced by thick glass mirrors silvered on the back. Such prisms are used in periscopes instead of glass mirrors (see Figure 2.3), in prismatic compasses, and in all accurate instruments which are required to produce clear images.

Advantage is taken of this property of total internal reflection in the erecting prisms shown in Figure 3.8. In both cases the light ray strikes the glass–air surface at angles greater than the critical angle. How fortunate that the critical angle for glass is 42°! Both prisms are often utilised in the construction of optical apparatus. A pair of erecting and reversing prisms are used to reduce the distance between the eye piece and the object glass for each eye in binoculars and therefore we call them prism or *prismatic binoculars*. Certain types of projectors require erecting prisms attached to them to invert the light beam before forward projection on to the screen.

The brilliant sparkle of diamonds is due to this same total internal reflection. They are cut so that nearly all the rays of light inside them striking upwards on sides below one set of facets, called the 'girdle', do so at angles greater than the critical angle of a diamond. These rays after several reflections inside emerge by the top facets to give plenty of sparkle.

Questions on chapter 3

1. A coin placed at the bottom of a white sink is just out of sight because the side of the sink obstructs the view of an observer. Water is gradually poured into the sink from a tap and the observer slowly begins to see the coin appear. Draw two diagrams to illustrate how this is possible.

2. Why do objects seen through the air above hot rocks or earth appear to quiver? Why has water in which ice or sugar is being dissolved a streaky appearance?

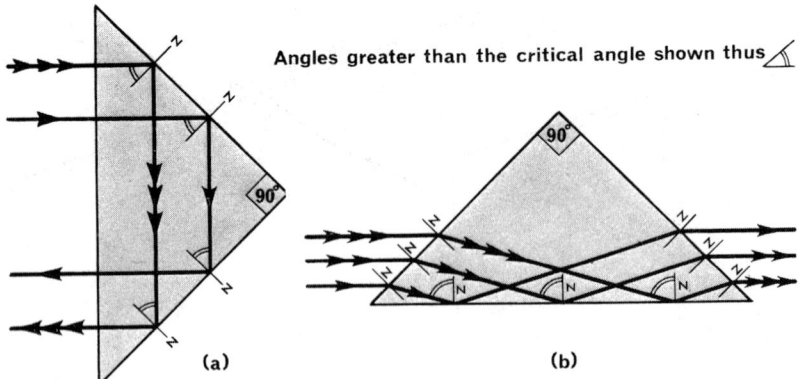

Figure 3.8. Two erecting prisms: (*a*) reverses the direction of the rays of light at the same time, and (*b*) allows the rays to continue in a forward direction. Both are right-angled isosceles glass prisms.

3. Explain with diagrams why a swimming pool appears to be shallower than it is in reality when it is looked at from above. Why does the pool appear to be still shallower when viewed from a distance?

4. What is meant by the refraction of light? State the law relating to the angles of incidence and refraction. Draw a diagram to show how the spectrum of white light may be formed by refraction through a glass prism. Mark on the diagram the angle of the prism and the angle of deviation for any one colour.

5. Why do stars twinkle?

6. Explain where you should aim to spear a motionless fish in a pond. Should you aim a little ahead or a little behind its apparent position or should you aim directly at the fish?

7. The apparent position of the sun is not its true position at sunrise or at sunset because refraction takes place as the light of the sun passes through the atmosphere. In fact, we see the sun for some minutes at night after it is really below the horizon. Why is this? At what time of the day will the true position and the apparent position be most nearly the same?

8. Draw diagrams to show what is meant by (a) critical angle, and (b) total internal reflection. Show how a triangular prism may be used to invert a beam of light.

9. Rays of light in a tank of water strike its surface with the air so that the angles of incidence are (a) 0°, (b) between 0° and the critical angle, (c) equal to the critical angle, and (d) greater than the critical angle. Draw diagrams of these four cases and show what happens to the light in each case.

10. Why is it possible to use 45° glass prisms as reflectors in binoculars? What advantage have these prisms over mirrors?

11. Explain why if you hold a glass of water with a spoon in it a little above the level of the eye, and look upwards at the under-surface of the water, you will find that you are unable to see that part of the spoon which is above the water.

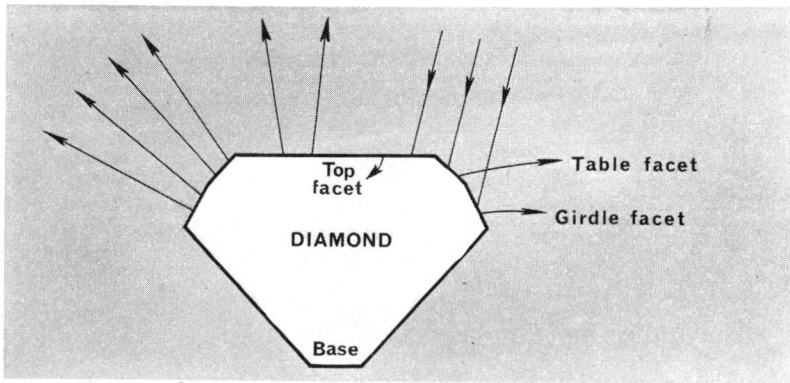

Figure 3.9. Why a cut diamond sparkles. Light from a bright source enters the three faces as shown on the right of the diagram. These three parallel rays after reflections and refraction leave in many directions from the three faces shown on the left. The same thing can happen on the other side producing 'sparkle' in very many different directions.

C

4. Dispersion and colour

Dispersion. A ray of white light sent through a *triangular prism* emerges in a new direction—it has been bent and the bending is called *deviation*. At the same time it will be noticed that the ray is spread out more than the original ray. This spreading out is called *dispersion* and if the ray is then examined very carefully it will be found to be *coloured*. It was in the year 1660 that this observation was first made by Sir Isaac Newton. The amount of dispersion is very small and can easily be overlooked and ignored altogether. The coloured patch of light that these rays form on a screen is called a *spectrum*.

Newton devised two experiments that you can easily repeat to show the reverse process—the production of white light from lights representing all the colours of the spectrum.

He recombined the colours of the spectrum by inserting a lens to converge the coloured rays to the same place on the screen. The same effect can also be produced by using instead of the lens another prism placed the reverse way up and by the side of the first prism.

His other experiment uses a principle that we call the *persistence of vision*. He painted the colours of the spectrum in the correct proportions on a circular disc. He then spun the disc in a bright light so rapidly that the eye believed it saw all the colours at the same time. This is because the retina of the eye continues to be influenced by one colour for about $\frac{1}{24}$ of a second, and during this time all the other colours have been able to add their effect one after the other on to the retina. Each colour thus persists continuously and the eye experiences the sensation of white light all the time the disc is spinning rapidly.

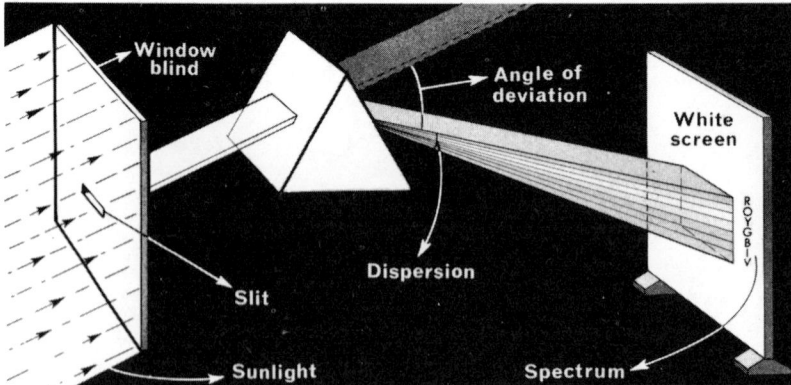

Figure 4.1. Newton's experiment to observe the refraction of light through a prism. The light is deviated and dispersed. The colours due to dispersion overlap one another and the red is seen to be least deviated and the violet most deviated. In this diagram the dispersion is exaggerated and the sunlight is reflected upwards towards the slit.

Spectra. It is possible to enlarge the dispersion that takes place when white light is deviated in passing through a prism. It is also nearly possible to separate the different colours from one another and to focus each one of them clearly on a screen. The essential procedure is to pass a narrow beam of parallel rays of white light through an equiangular prism so that the deviation produced is the minimum possible, and then to focus the coloured parallel rays. The best kind of prism to use is a hollow glass one filled with carbon disulphide, but solid flint glass and perspex prisms give good spectra. The coloured images produced in this way form a *pure spectrum*. Newton picked out as the most prominent colours in the spectrum—red, orange, yellow, green, blue, indigo, and violet.

If we use an instrument, called a spectroscope, built with lenses and a prism as shown in Figure 4.4, we can examine the spectra of various elements. Incandescent solids are found to give a continuous spectrum of visible light from red to violet. Look at the continuous spectrum in Figure 4.5.

Every different band of colour corresponds to certain light waves having its own particular band of wavelengths. In fact, a scale of wavelengths can be constructed for the whole of the visible spectrum and on into the infra-red and ultra-violet regions. It is convenient to use wavelengths measured in nano-metres for the scale because these are very very small. One thousand million of them equal 1 metre or 1 nm $= 10^{-9}$ m.

As each element has its own *bright line spectrum*, different from that of every other element, it is possible to detect the nature of a substance even if the amount available is extremely small. All one has to do is to turn the small amount of the substance into an incandescent vapour either by heating it in a flame or by enclosing it in an electrical discharge tube, and then to examine its spectrum.

By allowing light from the sun to enter a spectroscope the *solar spectrum* can be observed. This is a *continuous spectrum* due to the very hot incandescent core of the sun, but it is crossed by *black lines* due to the spectra of the elements in the cooler incandescent vapours of the corona. These black lines are not really

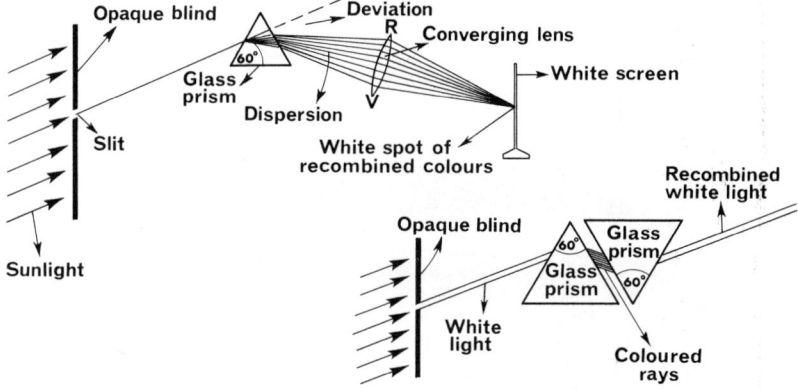

Figure 4.2. Two methods of combining the coloured rays of the spectrum to reform white light.

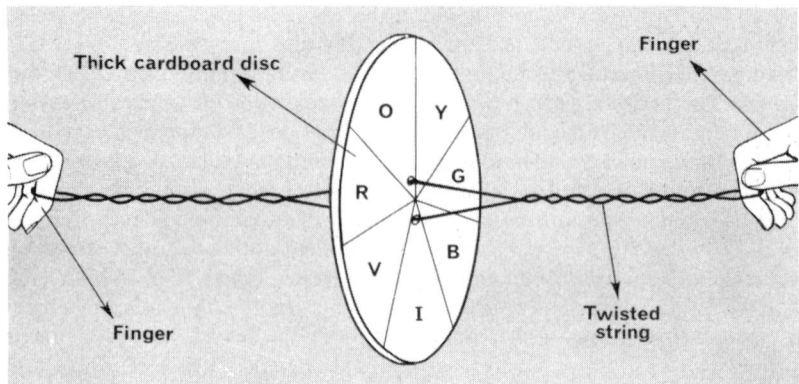

Figure 4.3. One way of spinning the circular disc so that each colour of the spectrum is seen by the eye in rapid succession. Is a pure white light observed?

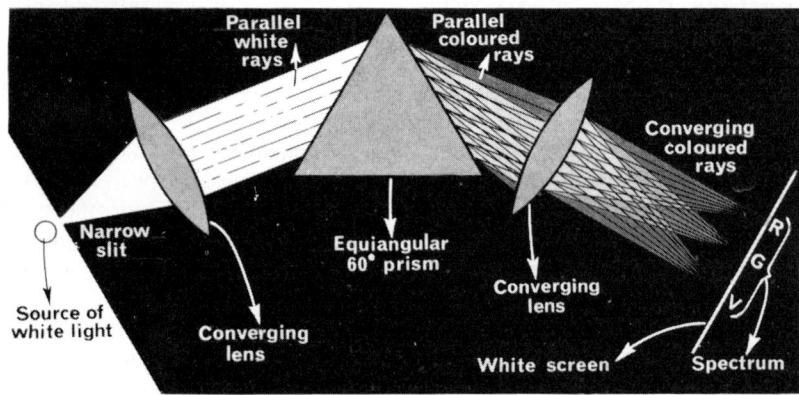

Figure 4.4. The formation of a pure spectrum. One converging lens provides the prism with parallel rays and the other converges the parallel coloured rays to form sharp coloured images of the slit on the screen.

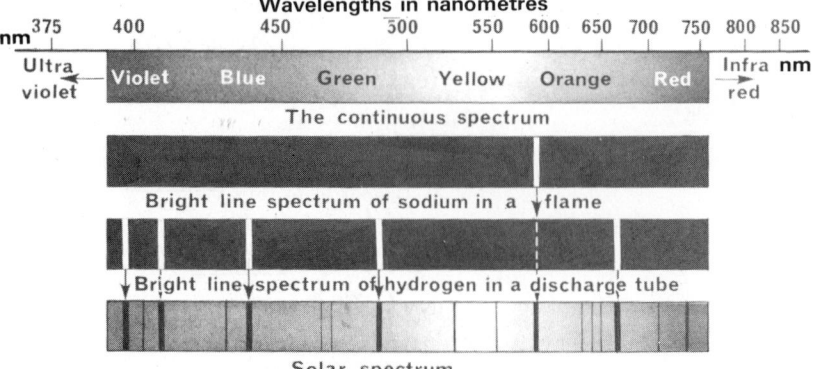

Figure 4.5. Types of spectra. The scale of these four spectra is marked at the top. Note how some of the dark lines in the solar spectrum correspond to the bright lines in the spectra of sodium and hydrogen. The other dark lines in the solar spectrum are not identified in these diagrams but belong to some other elements.

black but are not so brilliant as the continuous spectrum that forms the background so they appear by contrast to be black. If we examine these black lines we find they correspond to the lines of the spectra of the elements known on earth. This tells us that these elements are also present in the vapours of the sun's corona. However, some black lines were found in the solar spectrum that did not correspond to any bright lines we knew on earth, so we called the element responsible for these *helium* after the Greek word helios—sun. Years later we found helium on earth—it is inert, non-inflammable, and is used to fill airships and balloons because, except for hydrogen, it is the lightest gas known.

Primary and secondary colours. From the observations so far described we learn that white light can be produced by lights of all the colours of the spectrum put together. Moreover, if three of the prominent colours *red*, *green*, and *blue* shine as lights upon a screen the same colour results as if a white light had been shining on the screen. These three are called the *primary colours*. If only two of the primary coloured lights shine on a screen we get what we call a secondary coloured light. These *secondary colours* are formed thus:
1. Red plus green primaries give *yellow* secondary.
2. Green plus blue primaries give *peacock blue* secondary.
3. Blue plus red primaries give *magenta* secondary.

You can easily perform an interesting series of experiments to see these secondary colours by setting up the apparatus as shown in Figure 4.7. By erecting an opaque disc between the lanterns and the screen a colourful pattern of overlapping circles is formed that shows clearly how pairs of complementary colours together make up white light.

Complementary colours. It is an easy matter to find the complementary colour of any particular colour. Cut out a cross of the chosen colour and place it on a white background. Illuminate it strongly, shielding the source of light from

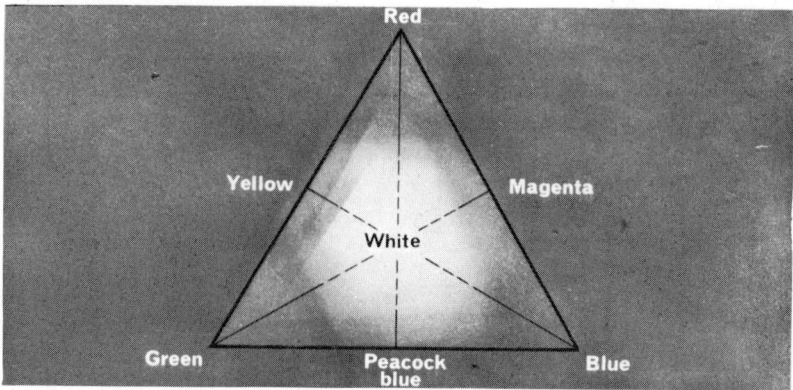

Figure 4.6. This triangle and its medians represent the main principles of colour combinations. Red, green, and blue are the primary colours; yellow, peacock blue, and magenta are the secondary colours; red and peacock blue, green and magenta, and blue and yellow are complementary pairs of colours. Why can the centre be represented by the total colour white?

your eyes. Stare at some single spot on the cross until your eyes become tired, then quickly pull the cross away whilst your eyes continue to stare at the white background. You will then see for a few seconds a cross in the colour which is complementary to the colour of the original cross.

How is the colour of an object determined? When we see a red object in daylight our eyes respond to what we call red light. How does this happen? White light strikes the object and its paint or pigment absorbs all the colours of the white light except red and this it reflects. If we look at the same object in a beam of red light only, then again red light is reflected and we still see the object as red. No other colours are absorbed by the object in this case as no other colours fall on it. In green light only, all the green is absorbed by the red object and no colour at all is reflected, hence the object appears to be black—the absence of colour. In a secondary yellow light the object reflects the red part of the yellow and absorbs the green part, so once again the object appears red. Thus whatever the colour of the light that falls on the *red* object it appears to be either *red* or *black* and nothing else. If the incident light that falls on it has red in it—the object is red, and if not the object is black.

Now consider a *yellow* object. It can reflect red and green light, or red only, or green only, or no light at all. Thus a yellow object can appear to be *yellow*, *red*, *green*, or *black* according to the colour of the light falling on it.

You can observe the colours of various coloured objects in primary and secondary coloured lights by using the colour lanterns described in Figure 4.7.

Thus the colour of an object depends on the colour of the light that falls on it and on the colours that it does not absorb.

An ordinary filament type electric lamp gives out light that has more red, orange, and yellow in it than daylight. Therefore it is impossible to match by filament lamps two materials for use in daylight if they have different colouring paints or pigments on them. That is why many shops have now fitted special lamps.

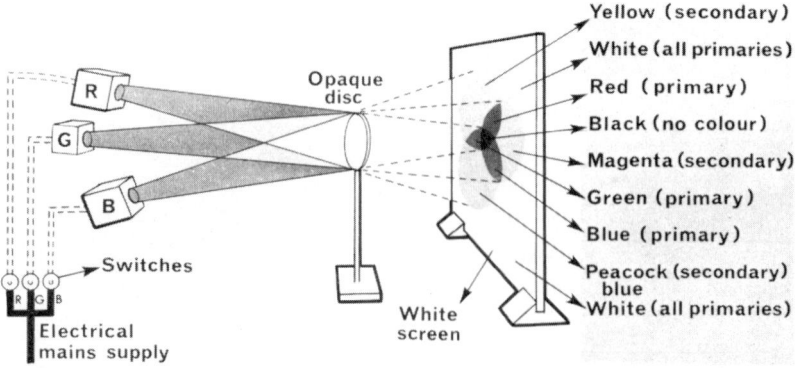

Figure 4.7. How to set up overlapping coloured shadows to illustrate primary and secondary colours. R, G, and B are three lanterns, floodlights, or torches on which red, green, and blue gelatine filter sheets have been fixed. These lanterns are mounted to form a triangle—G is farthest away and R is above the other two. They are controlled by three switches. Which pairs of colours are complementary?

Figure 4.8. This boy is finding the complementary colour of red. What would happen if he had a yellow cross on a blue background and the whole design were suddenly replaced by a white sheet of paper?

	First Transparent Layer	Second Transparent Layer	Third Transparent Layer	Transmitted Colour of Transparency
→	Magenta	Yellow	Peacock Blue	None (Black)
→	Magenta	Yellow	—	Red
Incident	—	Yellow	Peacock Blue	Green
White	Magenta	—	Peacock Blue	Blue
Light	Magenta	—	—	Magenta
→	—	Yellow	—	Yellow
→	—	—	Peacock Blue	Peacock Blue
→	—	—	—	White

Figure 4.9. A guide to the colours formed in the subtractive process of colour photography. White light passes through three coloured transparent layers to transmit eventually the colours marked in the last column.

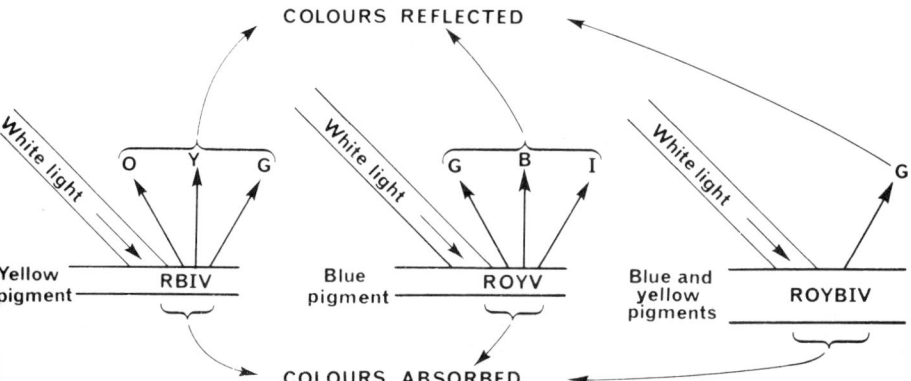

Figure 4.10. How a mixture of yellow (Y) and blue (B) pigments reflect green (G) light. The usual capitals ROYGBIV are used for the colours of the spectrum.

Colour transparencies. Suppose we covered a sheet of transparent film with a transparent layer of a pure secondary yellow dye and over it placed a transparent layer of a pure primary blue dye, what would be the resulting colour of the transmitted light? Because one layer of dye would absorb the colours that the other layer transmitted there would be a total absence of colour which we call black. But if we used with the transparent layer of a pure secondary yellow dye a transparent layer of a pure secondary peacock blue dye we should find that the colour common to both is transmitted and that it is green.

Therefore if white light is passed through a film on which has been deposited three transparent layers of the three secondary coloured dyes it is possible to transmit a variety of colours.

This is one of the processes in common use in colour photography. It is known as a subtractive process because each layer absorbs some section of the incident white light and transmits what remains.

Colours by pigments. A pigment is a coloured substance which, like paint, when applied to a surface, colours that surface. Pigments are prepared suitable for direct application and they are obtainable in most of the colours that the painter or artist requires. However, it is necessary sometimes to mix pigments to obtain a desired colour or shade. It is almost impossible chemically to obtain a pigment that will reflect a pure primary or secondary colour. For example, a yellow pigment will reflect a little orange and a little green at the same time as much yellow, and it will absorb the colours at the blue end of the spectrum. A blue pigment reflects not only much blue but a little green and indigo, and absorbs the colours at the red end of the spectrum. When yellow and blue pigments are mixed and applied to a white surface the resultant colour is that which is not strongly absorbed by either pigment. Therefore the colour most strongly reflected by the mixture of these two pigments is green.

Questions on chapter 4

1. Draw a diagram to show how a prism can be used to produce a spectrum from sunlight. Name the colours in their correct order. Explain what you would see when only red light falls on (a) a white wall, (b) a blue dress, and (c) a red and white striped football shirt.

2. A wide parallel beam of white light is passed through a triangular glass prism. Show clearly, on a large diagram, what happens to the beam of light. How would you (a) produce such a beam of parallel light, and (b) focus the emergent light on to a screen?

Explain why a blue object looks black when seen in the light from a street sodium lamp.

3. (a) Draw a diagram of the apparatus you would use to produce an ordinary spectrum, labelling the blue and the red ends of the spectrum.

(b) Explain why a piece of material appears a certain colour, e.g. red, in daylight.

(c) Explain why very little light can pass through a combination of a red and a blue filter.

(d) What are the three primary colours in light and what would be the effect of shining all three together on to the same area of a white screen?

4. Draw a diagram to illustrate the production of a pure spectrum. How would you show that the colours produced were pure? Explain why a white object looks nearly black when seen through sheets of red and blue glass held together.

5. A pure spectrum is thrown onto a white screen. What would be the shape and the colour of the spectrum left on the screen if a sheet of (a) yellow glass, (b) green glass, and (c) magenta glass is placed in the path of the light?

6. What colours would a bunch of red and yellow tulips appear if seen through a sheet of (a) red glass, (b) green glass, and (c) blue glass?

7. What is meant by 'complementary colours'? Yellow and blue lights are complementary. What will happen to the colour of a white screen illuminated only by yellow light if a beam of blue light falls onto it at the same time?

5. Lenses

Most optical instruments make use of lenses in some form or other—cameras, projectors, microscopes, telescopes, stereoscopes, and spectroscopes. All lenses are of two main types—*converging* and *diverging* according to the way they bend parallel beams of light.

A *converging lens* can be considered to consist of a number of thin triangular-shaped prisms placed one on the top of the other. It is then easy to understand how parallel rays of light striking a lens of this nature can be made to converge together to a single point called the *focus* or *focal point*.

By arranging the prisms the other way round a diverging lens is formed that diverges parallel rays as if they all came *from* a focus. This is a virtual focus because the rays do not actually pass through that point, but only appear to do so.

Images can be formed by lenses; they are like those formed by mirrors. There are, as in the case of mirrors, two rays that determine the construction of the image by drawing—these give the position, size, and nature of the image. Once the image has been located by drawing these rays it is fairly easy to trace the path of a beam of light from an object to an eye. One of these two important rays travels from the object parallel to the principal axis and after passing through the lens either goes through the focus (converging lens), or appears to have come from the focus (diverging lens). The other important ray travels from the object to the centre of the lens and continues in a straight line. Where these rays cross or appear to have crossed is the position of the real or virtual image.

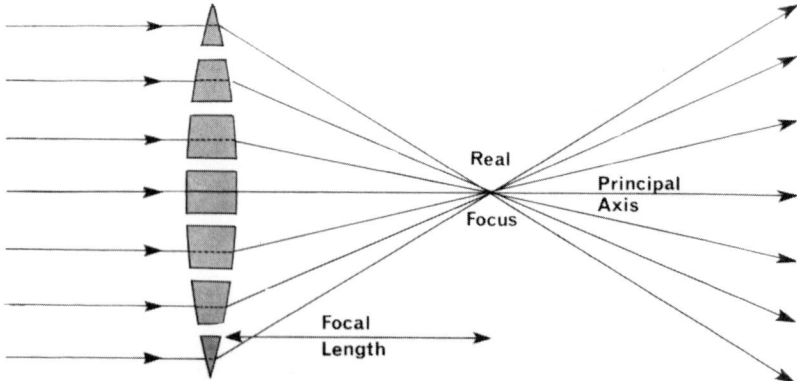

Figure 5.1. A converging lens is composed of an infinite number of prisms placed together as in this diagram until the two surfaces become smooth. This is a double-convex converging lens because both surfaces are convex.

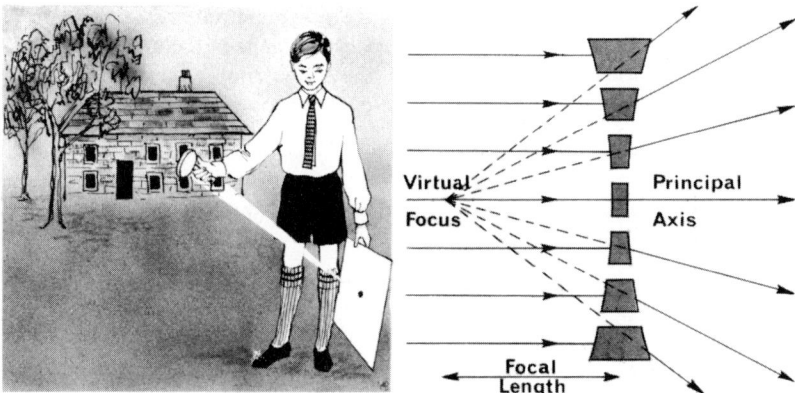

Figure 5.2. This boy is turning a converging lens towards the sun. On the paper a small spot of light is formed by the refraction of the parallel rays from the sun to the focus of the lens. The rays of heat concentrate there also and soon cause the paper to burn. The distance of the centre of the lens from the paper is the *focal length* of the lens.

Figure 5.3. A diverging lens shown in sections. Each section is part of a triangular prism. A diverging lens made in this way is a double-concave diverging lens.

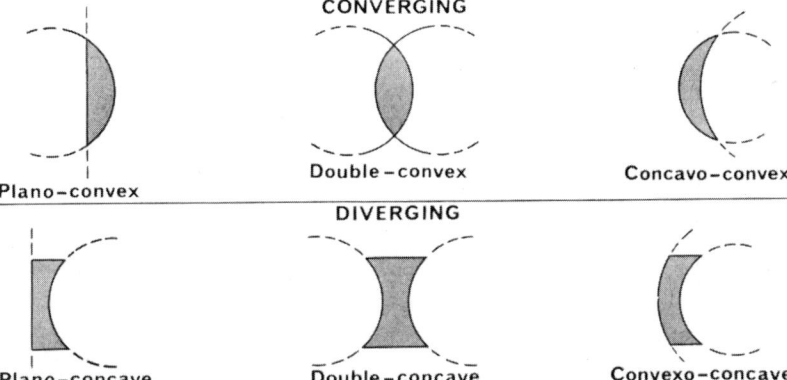

Figure 5.4. Converging and diverging lenses showing how their surfaces determine their names.

The table on pages 34 and 35 records the various situations in the case of the converging lens and some of the uses that are made of them. F denotes the position of the focus or focal point of the lens. 2F denotes a point on the axis situated twice as far from the lens as the focal point.

Position of Object	Construction of Image

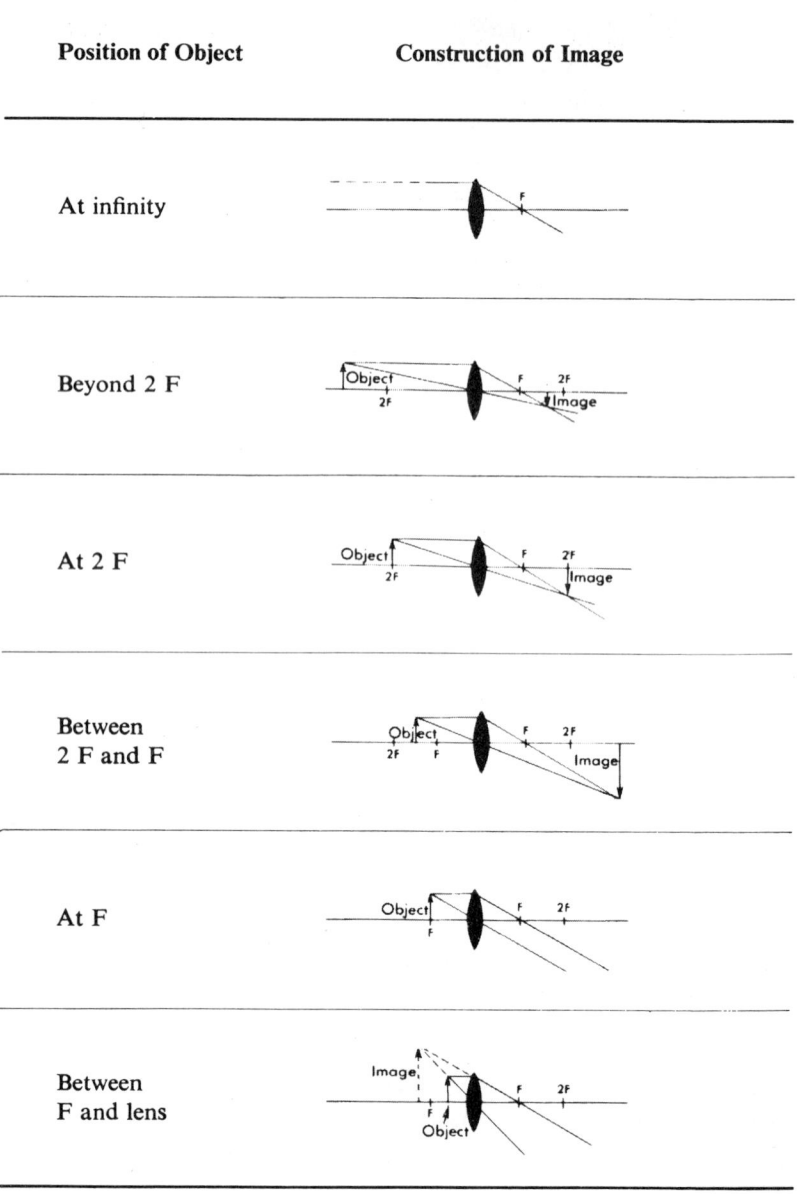

At infinity

Beyond 2 F

At 2 F

Between 2 F and F

At F

Between F and lens

| Image | | | Practical Uses |
Position	Relative Size	Nature	
At F	Very small	—	Burning glass. Telescope objective.
Between F and 2 F	Smaller	Inverted Real	The eye. Camera used for distant objects.
At 2 F	Same	Inverted Real	Copying camera used for 1 to 1 or life-size photographs.
Beyond 2 F	Larger	Inverted Real	Photographic enlarger. Projectors of all kinds. Microscope objective.
At infinity	—	No image— parallel beam	Spotlights. Light beams from lighthouses and ships.
Same side as object and further away from lens	Larger	Erect Virtual	Simple microscope or magnifying glass.

The human eye may be considered to be a spherical chamber with its front transparent and more strongly curved than the remainder of the eyeball. This front portion is called the *cornea*; behind it is a jelly-like substance and then a *converging lens* made of some fairly hard substance. At the back of the eye on the inside there is the *retina*, that is sensitive to light, and on which the image of an object is formed. The converging lens is able to focus the image from objects at various distances onto the retina by the action of the *ciliary muscles* altering the *curvature* of the surfaces of the lens. This process is known as *accommodation*. Normally the extent of this accommodation is limited and the eye cannot see distinctly objects much closer than about 25 cm. The *optic nerve*, carrying the messages from the sensitive retina to the brain, leaves from the back of the eyeball at a position known as the *blind spot*, so called because the eye cannot see an image formed there. The *yellow spot*, close to the blind spot, is the most sensitive section of the retina and it is on this spot that you manage to place the centre of the image that you wish to see most clearly. The *iris*, differently coloured in different people is situated between the cornea and the lens. It controls the quantity of light entering the eye by opening and closing the hole in its centre.

In the case of a *diverging lens* no matter where the object is, the image is always smaller, erect, and virtual. These lenses are used in spectacles to assist a short-sighted eye. They are also used in the construction of a pair of opera glasses.

Opticians and others speak of the *power* of a lens. A converging lens that can bend parallel rays of light to a focus a short distance away is more powerful (and at the same time much fatter) than one that can only bring them to a focus a greater distance away. The power of a lens is therefore defined as the reciprocal of its focal length when that is measured in metres. This power is measured in *dioptres*, so that if you are wearing $+\frac{1}{2}$ dioptre lenses in your spectacles the focal length of those lenses is 2 metres. The $+$ sign denotes that the lenses are converging ones.

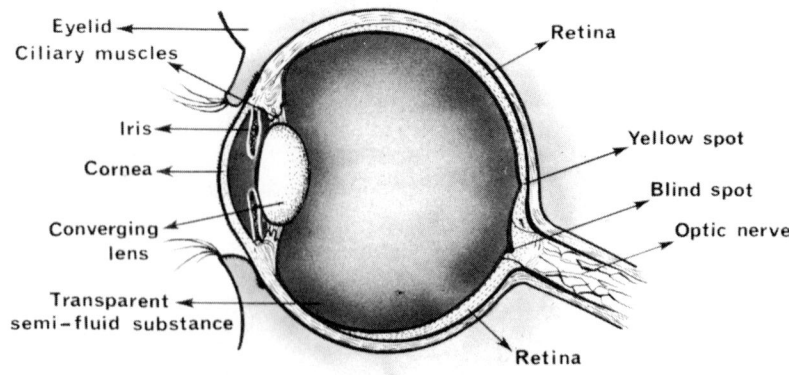

Figure 5.5. The human eye.

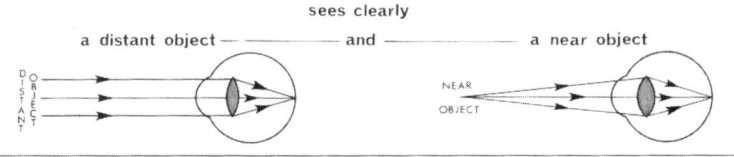

NORMAL EYE
sees clearly

a distant object ——————— and ——————— a near object

LONG SIGHTED EYE
sees clearly a distant object but only with the help of a lens a near object

Figure 5.6. The eyeball of a long-sighted eye is shorter than that of a normal eye and the muscles of the eye lens cannot make it fat enough to converge rays from a near object on to its retina. This diagram shows how an additional 'positive' converging lens enables the eye lens to correct this fault.

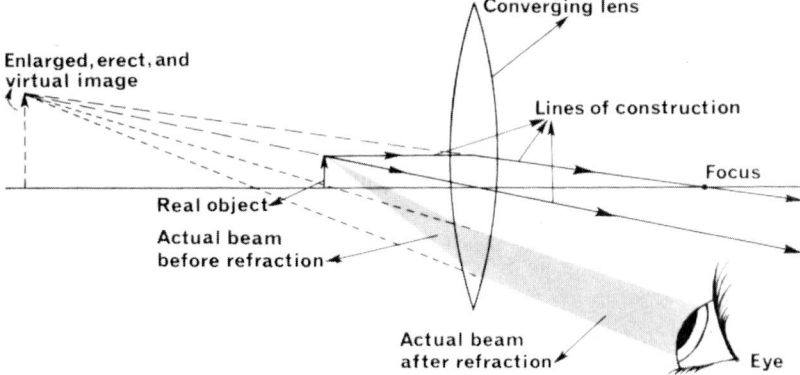

Figure 5.7. How an eye sees the greatly enlarged image of an object through a simple microscope or magnifying glass. The lens is put close to the object (less than its focal length away) and the eye is brought up so that it is at the position of minimum distance of distinct vision from the image. The beam from one point only on the object is drawn.

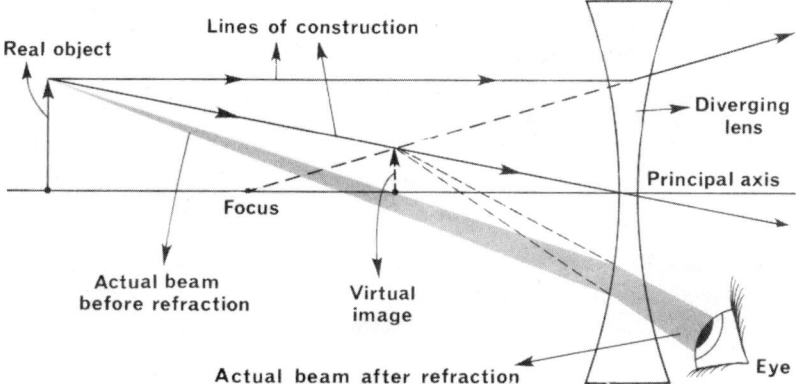

Figure 5.8. The construction of a virtual image of an object formed by a diverging lens. The eye sees this image by looking through the lens towards the object.

Projectors. The focal length of a lens must always be considered when we want to buy a lens for a particular purpose. For instance, long focal length lenses in projectors give narrow beams of light suitable for long rooms, whereas short focal length lenses give beams that spread out quickly, so that they can be used for short throws in small rooms.

Two types of projector are in common use to throw an enlarged image of an object on to a distant screen. Both types use converging lenses for projection. These projection lenses are normally compound lenses put together in one cylindrical frame so as to give the effect of one converging lens.

They are the *episcope* (epi means upon) so named because the light is thrown upon the object and then projected, and the *diascope* (dia means through) because the light is passed through the object and then projected. In both these projectors the object is placed slightly further away from the projection lens than the focus of the lens. In both projectors large converging lenses collect the light from powerful electric lamps and distribute it evenly onto the object. Both sources of light have concave mirrors arranged to collect the light travelling away from the object. These mirrors reflect this light forward to join with the forward travelling light passing through the converging lenses. The two converging lenses in the diascope form what is known as a condenser.

The powerful lamps in figures 5.10 and 5.11 are of the hot filament type. Recently a new type of lamp called the quartz iodine lamp has been fitted to some makes of projector. In this lamp the evaporated tungsten atoms of the filament combine with the iodine atoms near the cool quartz envelope. The tungsten iodide molecules so formed are then broken up by the intense heat of the filament and the tungsten atoms are deposited on the filament. This type of lamp gives a greater luminous intensity with less waste heat energy for the same electrical energy supplied from the mains.

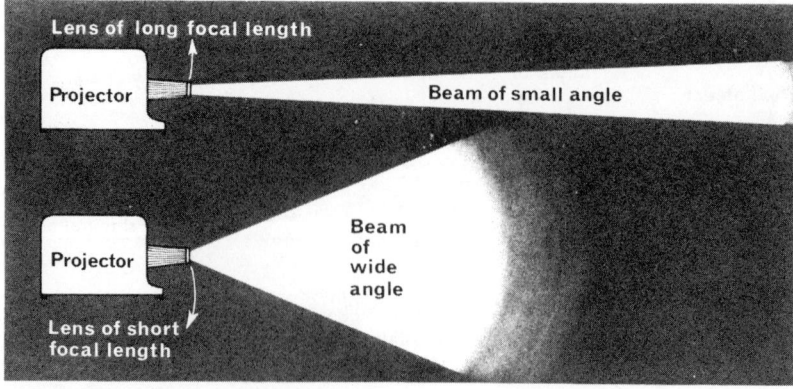

Figure 5.9. How to select a projector lens of the correct focal length in order to throw an image of a reasonable size on a screen in a long hall or a small room.

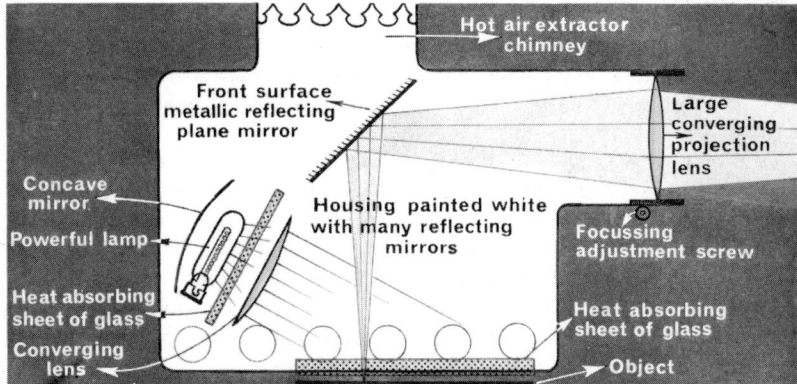

Figure 5.10. An episcope. The projection lens is very large so that it can gather as much as possible of the reflected light from the opaque object. Why is there a plane mirror in the optical system to invert laterally the light rays?

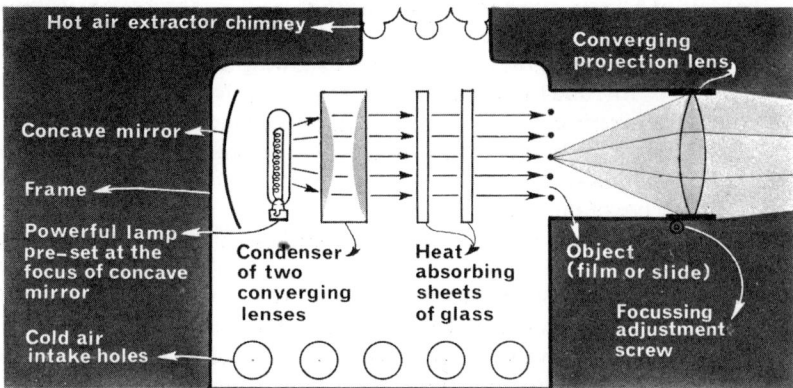

Figure 5.11. A diascope. Ciné-projectors, slide and filmstrip projectors, and optical lanterns are all of them diascopes. Two plano-convex converging lenses form the condenser that produces an evenly distributed beam of light on to the object from the light coming both directly from the lamp and by reflection from the concave mirror.

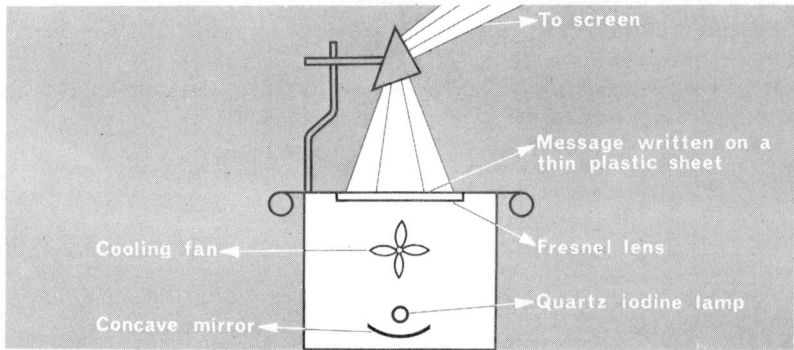

Figure 5.12. An overhead projector. This throws an image on to a vertical screen behind the operator.

D

Questions on chapter 5

1. State what construction lines can be used to find the position of the image formed by a converging lens of known focal length. Use this construction to show the formation of (a) a real image, and (b) a virtual image by such a lens.

2. Where would you place an object for it to form with a converging lens an image that is (a) real and diminished, (b) real and the same size, (c) virtual and magnified, and (d) virtual and the same size?

3. Find by scale drawing the size and position of an object 2 cm high placed 30 cm from a converging lens of focal length 10 cm.

4. A reading glass has a focal length of 12 cm. What is the magnification produced when it is held 10 cm away from a page of print? Make a drawing showing the lens, object, and image in these positions.

5. A converging lens of focal length 24 cm produces a real image three times as large as the object. What is the distance of the object from the lens?

6. A converging lens of focal length 6 cm is used to read the graduations on the scale of a screw gauge. The image must be erect and four times the size of the scale. Where should the lens be held and is the image real or virtual?

7. Draw a labelled diagram of a human eye showing the lens, cornea, retina, ciliary muscles, iris, and blind spot.

Describe an experiment to determine the focal length of a converging lens.

8. Explain with the help of diagrams how it is that we can see clearly either near or distant objects.

Explain the use of the iris in the human eye.

9. What happens in the eye when an object which is being viewed approaches the observer?

Draw diagrams to illustrate the defects of the eye known as 'long sight' and 'short sight' and how these may be corrected by the use of suitable lenses.

10. How do we see depth and judge distances? Should we be better able to to do this if our eyes were considerably further apart than they are now?

11. In a projection lantern, where is the slide placed with reference to the projection lens? How is the slide inserted in its holder and then placed in the projector?

12. Draw a labelled diagram of an optical projector.

Explain why it is necessary to have a condenser system and a high-power lamp in a projector.

If the linear magnification of a projection system is 50, what will be the width of the picture, formed on the screen, of a 35-mm slide?

6. Illumination

Most of us spend much of our time indoors where artificial lighting is required. A different intensity of illumination is needed for each of the tasks that have to be performed there. Sometimes the work is concerned with minute objects requiring great care and precision so that strong lamps close to these objects have to be provided. Thus it is important to know how much light is available, how it is distributed, and the quality of this light. The correct use of light is essential to our health and efficiency.

Good lighting must have sufficient intensity of illumination for the work to be done without strain, must be steady and even, and must not project glare into the eyes. At the same time the illumination should create shadows of reasonable contrast. Too much contrast causes eyestrain and discomfort, and too little leaves objects badly defined, so that one cannot distinguish them clearly from their backgrounds with the result that accidents can easily occur.

Photometry deals with the problems of measuring light. There are three quantities with which we shall be concerned: the *luminous intensity* of the source (I), the *luminous flux* or light flow from a source (Φ), and the *intensity of illumination* on a surface (E).

The luminous intensity of a source (I) is measured in candela (cd). There are standard electric lamps calibrated in *candela* and these can be used to determine the luminous intensity of another source as the boy is doing in Figure 6.2.

Of all the energy radiated from a luminous source only a certain percentage

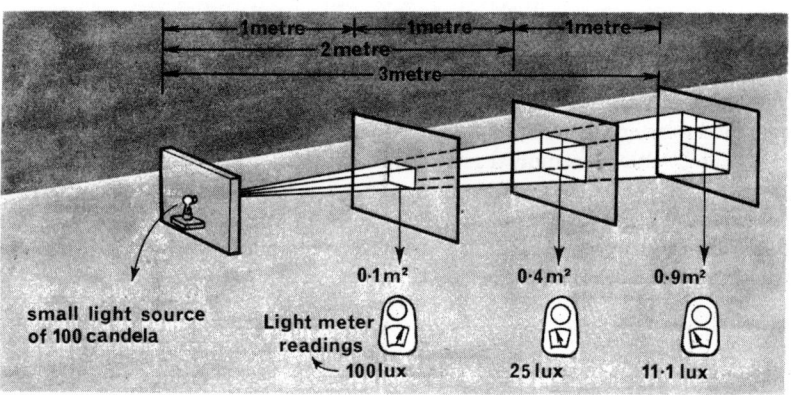

Figure 6.1. This shows how the illumination on a surface varies with its distance from the source. The inverse-square law applies only to small sources of light. The light-meter readings are taken on the surfaces when the meter is lying normally to the rays of light from the source.

41

is capable of producing a visual sensation. In some cases only 10% is what we call visual light. This is called the luminous flux (Φ) and is measured in *lumen*.

The intensity of illumination (E) is a measure of the density of luminous flux falling on a surface and is therefore luminous flux per unit area.

Hence
$$E = \frac{\Phi}{A}.$$

1 lumen of flux falling perpendicularly on 1 square metre of area produces an intensity of illumination of 1 *lux*.

The illumination on any surface depends not only upon the luminous intensity of the source but also upon the distance of the surface from the source. Of course, other factors can interfere with the illumination such as the presence of mirrors, reflecting surfaces, and lenses, but if these can be removed we can determine the relationship between illumination, luminous intensity, and distance.

If we set up the apparatus in Figure 6.1 we shall soon see with the aid of a light meter that the illumination decreases in proportion as the square of the distance increases. This can be explained by the fact that all the light passing through 0.1 m² at 1 m distance passes through 0.4 m² at 2 m distance and through 0.9 m² at 3 m distance. This is called the inverse-square law of illumination.

$$\text{Intensity of illumination in lux} = \frac{\text{Luminous Intensity of source in candela}}{(\text{Distance in metres})^2}$$

$$\text{or } E = \frac{I}{d^2}$$

Using the information given in Figure 6.1 we calculate as follows:

$$\text{Intensity of illumination at 1 m} = \frac{100}{1^2} = 100 \text{ lux.}$$

$$\text{Intensity of illumination at 2 m} = \frac{100}{2^2} = 25 \text{ lux.}$$

$$\text{Intensity of illumination at 3 m} = \frac{100}{3^2} = 11.1 \text{ lux.}$$

Photometers are used to compare the luminous intensities of sources of light. An easy photometer to make and use is the *grease spot* photometer, shown in Figure 6.2, which is simply a grease spot smeared in the middle of a piece of plain paper. When the illuminations on both sides of the paper are equal the grease spot becomes almost invisible. It is then necessary to measure the distances the sources of light are away from the photometer, d_s and d_x, and substitute these in the equation:

$$\frac{I_s}{d_s^2} = E = \frac{I_x}{d_x^2}$$

where I_s and I_x are the luminous intensities of S and X, and E is the intensity of illumination on both sides of the grease spot.

What are reasonable values for the illumination of our rooms taking into account all the reflections and absorptions due to walls, ceilings, doors, and

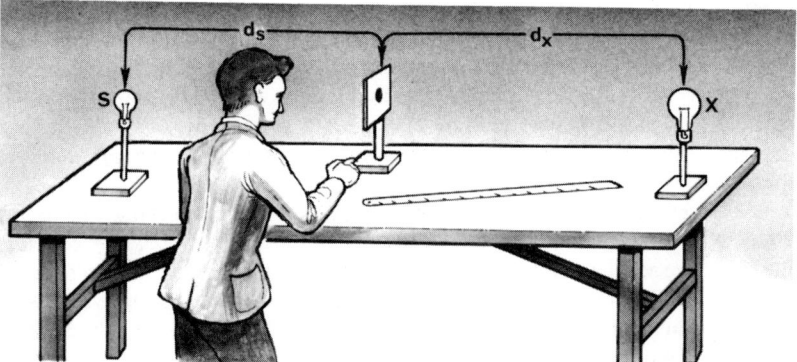

Figure 6.2. This boy is comparing the luminous intensities in candela of the lamp S and the lamp X. He is adjusting the 'grease spot' photometer until the grease spot disappears as seen from either side. He is working in a dark room so that no other light can interfere with his readings.

curtains? To help you to compare the values listed below, illumination in bright sunshine may reach 1 000 lux in England and 2 000 lux high up in the mountains, about 100 lux in the shade, and about 0.002 lux in the brightest moonlight.

Purpose	lux	Purpose	lux
Ordinary reading	1–200	Ironing and laundry work	1–200
Reading fine type	2–500	Corridors and stairs	20–50
Ordinary sewing	1–200	Living rooms	50–100
Prolonged sewing	2–500	Workshop benches	2–300
Sewing dark material	5–1000	Laboratory benches generally	2–300
Fine needlework	over 1000	Drawing office benches	2–300
Normal writing	1–200	Operating benches	over 1000
Normal kitchen work	50–100	Watch repairing	over 1000
Cooking and washing	1–200		

Luminous Efficacy of Lamps. The number of lumens radiated for every watt of electricity consumed by the source is known as the *luminous efficacy* of that source. Of course this depends on many factors such as the material of the filament, how it is wound, and the general design of the lamp. For example, the number of lumens per watt for the following lamps are approximately:

Vacuum type metal filament	80	Sodium vapour	120
Gas filled coiled coil metal filament	1500	Mercury vapour	50

Glare is one of the most common faults in our lighting schemes. The reflection of light on the surfaces in a room has an important effect not only on the actual illumination, which we have already discussed, but also upon glare. Painted walls and ceilings should have matt surfaces so that they can reflect a comfortable light by diffusion. Artificial light should have diffusing glass or light coloured plastic shades or bowls as antiglare devices. It is undesirable to illuminate shiny surfaces with naked lamps. Books, magazines, and tables often have shiny surfaces and they reflect images of a naked lamp, making reading difficult. The best reading surface is a matt paper that will diffuse the light from its surface and not form bright images by reflecting it like a mirror.

We sometimes try to see something that is almost in the direct line of a bright lamp. Not only does this produce eyestrain and discomfort, but by contrast the object appears to be badly illuminated as shown in Figure 6.3. In fact the direction of the source of light should be behind or nearly at right-angles to the direction needed to view the object. Any unshaded source of light close to the line of vision reduces the effectiveness of the actual illumination.

It is also important that the illumination of a book, for instance, should not be more than ten times the illumination of the background of the room in which the book is being read. It is thus necessary to illuminate correctly not only the book but also the room.

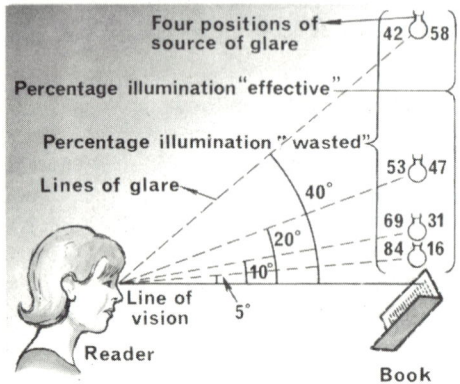
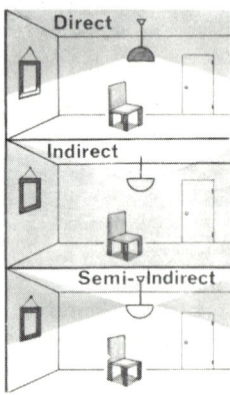

Figure 6.3. This diagram shows the effectiveness of a source of light in illuminating a book that is being read. It also shows how much light is wasted by producing glare and thus disturbing and closing the eyes of the reader. The angle formed at the eyes between the glare source and the book is marked. When this angle becomes small the book is almost unreadable.

Figure 6.4. Three types of lighting installation. Good illumination for working should be adequate, well diffused, free from glare, and evenly distributed.

Illumination. How should the artificial illumination of a room be planned? There is *direct* illumination that sends sometimes as much as half of the light from the source directly where it is needed. There is *indirect* illumination so arranged that all the light goes to some reflecting surface, such as the ceiling, and from there it is diffused evenly all over the room.

Direct illumination is the more economical of the two and is supplied either by naked lamps or by frosted lamps in diffusing bowls. Sometimes the lamps are placed behind frosted diffusing glass in a recess in the ceiling. Provided that the light is well diffused and its source properly placed this type of lighting can be satisfactory. Indirect illumination is more expensive to maintain because more candela are needed to provide the same illumination on any particular surface. At the same time indirect lighting produces the least annoying shadows and is more pleasant if the ceiling illumination is not excessively great.

The most practical type of illumination for ordinary purposes is a combination of these two types. This can be provided by fitting a translucent bowl below the lamp so that some light is thrown upwards to the ceiling whilst the remainder is diffused directly to the working surface. This should then give adequate illumination without glare, and a reasonable uniformity so that the working surface is not too bright or too dim in relation to the rest of the room.

Questions on chapter 6

1. Describe how to use a photometer. What law enables us to compare the luminous intensities of two light sources with a photometer?

2. Compare the different lighting systems for the illumination of a room. What are their characteristic features? How does the quality of the illumination vary in each method you describe? Which is the most economical system if only intensity of illumination and cost are considered?

3. What is meant by the term lux? For reading in comfort the intensity of illumination falling on the book should be 200 lux. How far above the book should a lamp of 200 candela be placed to obtain this illumination?

4. On a bright day in summer the illumination outside in the open country is 60 000 lux. Compare this with the illumination of a sheet of paper in a dark room placed 1 m below an electric lamp of 24 candela.

5. A 12 volt motor-car lamp placed 6 m from a screen gives the same intensity of illumination as a torch lamp placed 1 m away from the screen. Compare their luminous intensities.

6. A room is illuminated by an electric lamp placed inside a frosted glass bowl. The intensity of illumination it produces on a surface 2 m away from the bowl is the same as that produced by the same unshaded lamp placed 3 m from the surface. What percentage of the light given out by the lamp is cut off by the bowl?

7. If a 16 candela electric lamp at a distance of 10 m illuminates a surface to a particular degree of brightness, at what distance must a 36 candela electric lamp be placed from the surface to illuminate it to the same degree of brightness?

8. Four standard lamps of 1 candela luminous intensity each are placed very close together and 1 m from one side of the screen of a grease-spot photometer. How far must a 2 500 candela electric lamp be placed from the other side in order to cause the disappearance of the grease spot?

7. Movement

A body may move in many ways. It may move in a straight line, in a smooth curve, or back and forth; it may simply spin round some central axis, or it may move in an irregular way in any direction. In this chapter we shall only consider the movement of a body along a straight line.

Examine the graphs shown in Figure 7.1. The three trains cover the same distance in different times, or expressed in another way, each train during the interval of time covers a different distance. The speed of each train is shown by the slope of its graph—the steeper the slope the faster the train. The fastest train covers the greatest distance in any given interval of time. *Speed* can thus be defined as the distance travelled in a unit of time, or the rate at which a body travels through space, or the rate of change of position. The equation that represents this is:

$$\text{SPEED} = \frac{\text{DISTANCE}}{\text{TIME}}$$

Speed can be measured in a number of different units such as metres per second, or kilometres per hour, and these are written as cm s^{-1} or km h^{-1} but some instruments used in a car for example may be marked k.p.h. or km.p.h. In our science we shall always use the notation km h^{-1} which means kilometres per hour.

The definition of the speed of a body applies equally well to a body moving along a path that curves. For example, a motor-car travelling along a twisty road might maintain a speed of 50 kilometres per hour and if it did this it

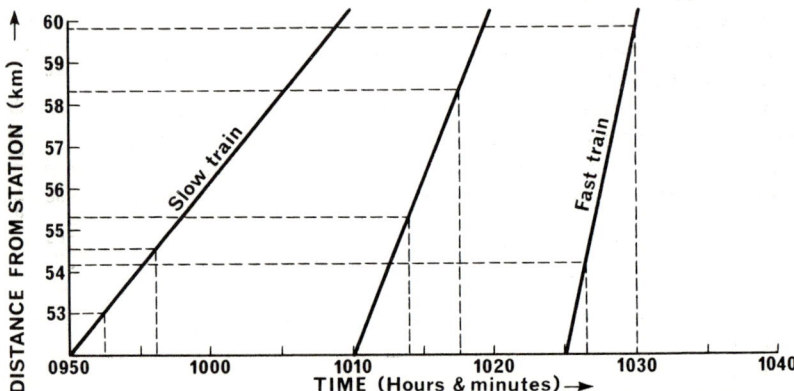

Figure 7.1. The distance–time graphs of three trains travelling over a small section of a railway system. Each pair of vertical dotted lines is drawn to enclose equal time intervals and the corresponding pairs of dotted horizontal lines indicate the distances travelled by each train during the time interval.

46

would cover a distance of 50 km every hour. If, however, this same car travels along a straight road and continues to keep to that straight road then it is said to have a fixed direction of movement as well as a speed of 50 kilometres per hour. In other words it is then said to have a velocity of 50 kilometres per hour in a certain direction. This need to state direction is the only difference between speed and velocity. In this chapter when the word speed is used it could be replaced by the word velocity.

On a long journey by car one may travel a distance of 250 kilometres in 5 hours. This means that the speed for the journey is 50 kilometres per hour. Obviously the car does not travel all the time at 50 kilometres per hour, sometimes it is more and sometimes it is less. This 50 kilometres per hour represents the average speed for the journey. When the driver wishes to overtake another car he increases his speed and this rate of gain in speed is called *acceleration*. When he slows down again *retardation* takes place.

For the movement of a body two graphs can be drawn, one a distance-time graph and the other a speed-time graph. If the speed-time graph is a straight line inclined to the axis it means that the body is under the influence of a constant acceleration. This is the case with the examples chosen in Figures 7.2 and 7.3, for the attraction of gravity on *any* body near the surface of the earth gives it a constant acceleration called g. It is very difficult to detect any difference in the motion of two different heavy bodies when they are released from a great height. The acceleration g of two such bodies has been determined by accurate timing methods and found to be always the same at the same place. It is very nearly 9.81 m s^{-1} every second or 9.81 m s^{-2}. This value of g varies slightly from place to place according to its particular elevation and latitude, and to the density of the rock below the surface of the earth.

The speed of the three trains during the small section of their journeys shown by the distance-time graphs in Figure 7.1 remains constant. But if a larger section of their journeys is examined it will be seen that the speed of each train changes as it proceeds on its way. In fact, each train starts from rest and slowly

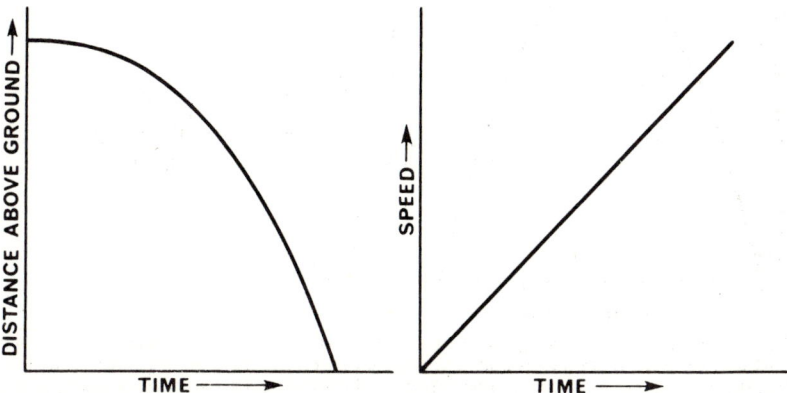

Figure 7.2. A distance-time graph and a speed-time graph for a heavy stone falling freely in space and neglecting any retardation due to the air.

increases its speed until its maximum speed is reached. Then this speed is maintained for the greater part of the journey, and finally its speed drops again to zero. A simple graph showing how the speed of a train changes during a journey with respect to the time is shown in Figure 7.4. The rate of change of speed (the acceleration) is calculated from the slope of the graph, and this calculation can be made at any point on the graph by drawing a tangent to the graph at that point. During the period of acceleration at the beginning of the journey the slope of the graph is in one direction and during retardation towards the end of the journey the slope is reversed. The steeper the slope the greater the acceleration or retardation. The equation that represents this is:

$$\text{ACCELERATION} = \frac{\text{CHANGE OF SPEED}}{\text{TIME}}$$

Acceleration can be expressed in several ways: kilometres per second every second or kilometres per second per second or simply as km s^{-2}.

The braking power of the brakes of a motor-car is measured by the retardation (negative acceleration) produced when the brakes are applied. This measurement is an essential part of any road test carried out on a motor-car. Safety belts for passengers are required so that the speed of the passenger can be retarded with the same value as the body of the car; otherwise the passenger may be thrown against the windscreen as he continues his forward movement during the time the car is braking. For ordinary motor-cars the maximum retardation is usually much greater than the maximum acceleration.

Equations are used to establish the relations between the distance, speed, and acceleration of a body.

When the speed of a body is constant the acceleration is zero, and

$$\text{SPEED} = \frac{\text{DISTANCE}}{\text{TIME}}, \quad \text{or} \quad \text{DISTANCE} = \text{SPEED} \times \text{TIME}$$

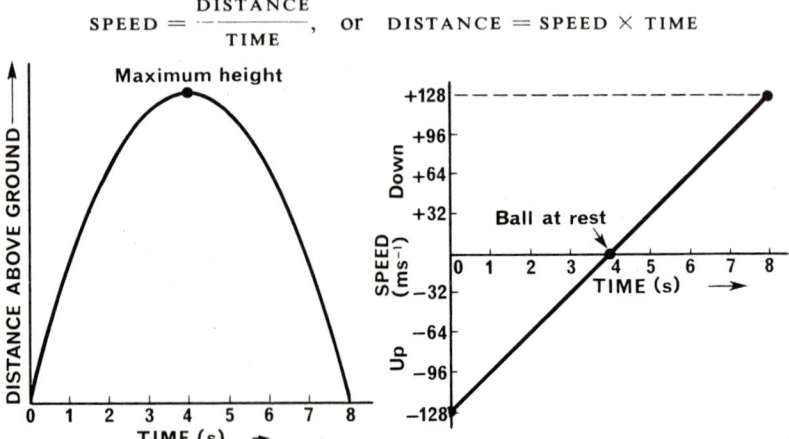

Figure 7.3. A distance-time graph and a velocity-time graph for a ball being thrown vertically upwards from the ground. The retardation due to the air is neglected, and therefore the ball reaches the ground again where its distance from the starting position is zero and its velocity is the same as its starting velocity but in the opposite direction. Note the scales used and that the downwards direction of the velocity is positive. When does it reach its highest position and when does it reach the ground again?

These equations can be written using the symbols u for the speed, s for the distance, and t for the time, thus:

$$u = \frac{s}{t}, \quad \text{or} \quad s = u \times t \quad . \quad \quad . \quad . \quad \quad (1)$$

Example 1. If a hiker covers 24 km in 8 hours, what is his average speed? If he walks at the same average speed during the next day for 10 hours, how far will he travel?

(a) $u = ?$, $s = 24$ km, and $t = 8$ hours.

$$\text{speed} = \frac{\text{distance}}{\text{time}},$$

$$\text{or } u = \frac{s}{t}, \quad . \quad \quad . \quad \quad . \quad \quad . \quad \quad . \quad (1)$$

$$\therefore u = \frac{24 \text{ km}}{8 \text{ hours}},$$

$\therefore$ His average speed = 3 kilometres per hour.

(b) $u = 3$ kilometres per hour, $s = ?$, and $t = 10$ hours.

$$\text{distance} = \text{speed} \times \text{time},$$

$$\text{or } s = u \times t, \quad . \quad \quad . \quad \quad . \quad \quad . \quad (1)$$

$$\therefore s = \frac{3 \text{ km}}{\text{hour}} \times 10 \text{ hours},$$

$\therefore$ Distance hiker travels = 30 kilometres.

When the speed of a body changes and there is an acceleration or a retardation the starting speed and the final speed will be different. In order to determine the acceleration or retardation the time taken for the speed to change must be known.

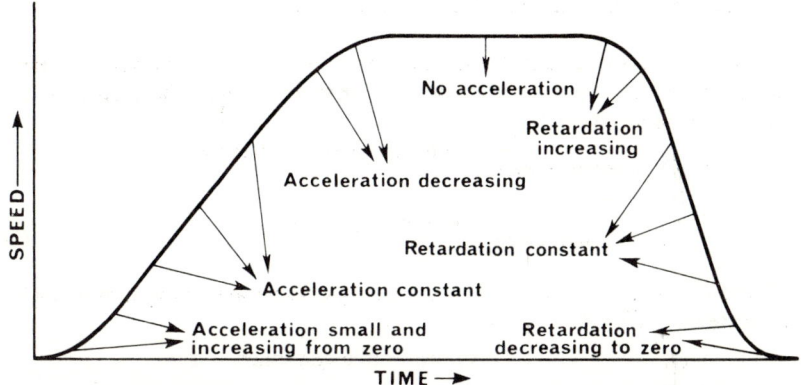

Figure 7.4. A speed–time graph for the whole journey of a train from start to finish. When the retardation is greater than its acceleration the brakes are probably in good order.

Example 2. A train starting from rest reaches a speed of 36 kilometres per hour in 100 seconds. What is the average acceleration of the train?

For convenience we shall calculate the next examples in metres and seconds.

In order to distinguish between the two different speeds we shall use the symbol u for the starting speed and v for the final speed. We shall use a for the acceleration.

$$u = 0 \text{ m s}^{-1}, v = 10 \text{ m s}^{-1}, a = ? \text{ and } t = 100 \text{ s}$$

Note that $36 \text{ km h}^{-1} = 10 \text{ m s}^{-1}$

$$\text{average acceleration} = \frac{\text{change of speed}}{\text{time}},$$

$$\text{or } a = \frac{v - u}{t}, \qquad . \qquad . \qquad . \qquad . \qquad (2)$$

$$\therefore a = 10 \text{ m s}^{-1} \times \frac{1}{100\text{s}},$$

$$\therefore \underline{\text{average acceleration of the train} = 10^{-1} \text{ m s}^{-2}.}$$

Example 3. A train running at 72 km h^{-1} starts to brake and after 25 seconds has slowed down to 36 km h^{-1}. What is the average retardation?

$$u = 20 \text{ m s}^{-1}, v = 10 \text{ m s}^{-1}, a = ?, \text{ and } t = 25 \text{ s}$$

Note that $72 \text{ km h}^{-1} = 20 \text{ m s}^{-1}$ and $36 \text{ km h}^{-1} = 10 \text{ m s}^{-1}$

$$\text{average acceleration} = \frac{\text{change of speed}}{\text{time}},$$

$$\therefore a = (10 - 20) \text{ m s}^{-1} \times \frac{1}{25 \text{ s}},$$

$$\therefore a = \frac{-10 \text{ m s}^{-1}}{25 \text{ s}}.$$

$$\therefore \underline{\text{average retardation of the train} = 0.4 \text{ m s}^{-2}.}$$

If the acceleration remains the same as shown in the section marked 'Acceleration constant' in Figure 7.4, then the average speed is simply the average of the starting and final speeds:

$$\text{average speed} = \frac{\text{starting speed} + \text{final speed}}{2}$$

$$\text{or average speed} = \frac{u + v}{2} \qquad . \qquad . \qquad . \qquad . \qquad (3)$$

The distance travelled by a body in uniformly accelerated motion can now be calculated by using the relation:

$$\text{distance} = \text{average speed} \times \text{time}$$

$$\text{or } s = \frac{u + v}{2} \times t . \qquad . \qquad . \qquad . \qquad (4)$$

The final speed of a body in any constant accelerated motion is simply the starting speed plus the speed gained during the acceleration of the body, and this is written:

$$\text{final speed} = \text{starting speed} + \text{acceleration} \times \text{time}$$

$$\text{or } v = u + a \times t. \quad . \quad . \quad . \quad . \quad . \quad (5)$$

Example 4. What is the speed of the train, described in Example 2, at a time 10 seconds after it started and how far has it then travelled?

$u = 0 \text{ m s}^{-1}, v = ?, s = ?, a = 10^{-1} \text{ m s}^{-2}, \text{ and } t = 10 \text{ s}$

$$v = u + at, \quad . \quad . \quad . \quad . \quad . \quad . \quad (5)$$

$$\therefore \ v = 0 + 10^{-1} \text{ m s}^{-2} \times 10 \text{ s},$$

$\therefore$ speed of the train after 10 seconds $= 1 \text{ m s}^{-1}$

$$s = \frac{u + v}{2} \times t, \quad . \quad . \quad . \quad . \quad . \quad (4)$$

$$\therefore \ s = \frac{0 + 1}{2} \text{ m s}^{-1} \times 10 \text{ s},$$

$\therefore$ distance travelled by the train

$$\text{in the first 10 seconds} = 5 \text{ metres.}$$

Example 5 How far does the train in Example 3 run during the braking?

$u = 72 \text{ km h}^{-1}, v = 36 \text{ km h}^{-1}, s = ?, \text{ and } t = 25 \text{ s}.$

Note that $72 \text{ km h}^{-1} = 20 \text{ m s}^{-1}$, and $36 \text{ km h}^{-1} = 10 \text{ m s}^{-1}$

$$s = \frac{u + v}{2} \times t \quad . \quad . \quad . \quad . \quad . \quad (4)$$

$$\therefore \ s = \frac{20 + 10}{2} \text{ m s}^{-1} \times 25 \text{ s},$$

$$\therefore \ s = 15 \times 25 \text{ m}.$$

$\therefore$ Distance the train travels

$$\text{during braking} = 375 \text{ m.}$$

Example 6. A stone is dropped down a deep well and is heard to splash in the water after 5 seconds. With what speed does the stone hit the water and how deep is the water below the top of the well, if the acceleration due to gravity is taken to be 9.8 m s^{-2} and the speed of sound is neglected?

$u = 0 \text{ m s}^{-1}, v = ?, s = ?, a = 9.8 \text{ m s}^{-2}, \text{ and } t = 5 \text{ s}.$

$$v = u \times a \times t, \quad . \quad . \quad . \quad (5)$$

$$\therefore \ v = 0 + 9.8 \text{ m s}^{-2} \times 5 \text{ s},$$

$\therefore$ speed the stone hits the water $= 39.0 \text{ m s}^{-1}.$

$$s = \frac{u + v}{2} \times t, \qquad \cdots \qquad (4)$$

$$\therefore\ s = \frac{0 + 39}{2}\ \text{m s}^{-1} \times 5\ \text{s},$$

$\therefore$ <u>depth of the well = 97.5 m.</u>

Example 7. A motor-car is tested for the efficiency of its brakes. It is driven along at a steady speed of 36 kilometres per hour and it comes to rest in 2 seconds after the brakes are firmly applied. A retardation less than 40% of g is considered poor, 40% to 55% is fair, 55% to 70% is good, and over 70% is excellent. What is the condition of the brakes of this car? How far does the car travel after its brakes are applied? Take g to be 9.8 metres per second every second, and assume the car is uniformly retarded by the brakes.

(a) $u = 36\ \text{km h}^{-1} = 10\ \text{m s}^{-1}$, $v = 0\ \text{m s}^{-1}$, $a = ?$, and $t = 2\ \text{s}$

$$v = u + at, \qquad \cdots \qquad (5)$$

$$\therefore\ 0 = 10\ \text{m s}^{-1} + a \times 2\ \text{s},$$

$$\therefore\ a = -5\ \text{m s}^{-1},$$

$\therefore$ the brakes of the car produce a retardation of $5\ \text{m s}^{-1}$.
$\therefore$ the retardation is 5/9.8 of the value of g, or 51% of g.

$\therefore$ <u>the brakes of the car are in 'fair' condition.</u>

(b) $u = 10\ \text{m s}^{-1}$, $v = 0\ \text{m s}^{-1}$, $s = ?$, and $t = 2$ s.

$$s = \frac{u + v}{2} \times t, \qquad \cdots \qquad (4)$$

$$\therefore\ s = \frac{10 + 0}{2}\ \text{m s}^{-1} \times 2\ \text{s},$$

$$\therefore\ s = 10\ \text{m}.$$

$\therefore$ <u>the car is brought to rest by the brakes during this test in 10 metres.</u>

Some useful equations concerning the movement of bodies are derived mathematically from the fundamental equations (4) and (5) :

(a) when starting from rest ($u = 0$).

$$v = at$$
$$s = \tfrac{1}{2}at^2$$
$$v^2 = 2as$$

(b) when the starting speed is u.

$$v = u + at$$
$$s = ut + \tfrac{1}{2}at^2$$
$$v^2 - u^2 = 2as$$

Questions on chapter 7

In all these questions the effects of friction and air resistance are neglected. Assume g to be 9.8 m s^{-2}.

1. Distinguish between speed and velocity. What is meant by uniform acceleration?

A stone is projected vertically upwards with a velocity of 49 m s^{-1}. Calculate how high it will rise and after what interval it will arrive again at its starting point.

2. Define (a) velocity, and (b) acceleration. State one unit in which each of these quantities may be measured.

A brick is dropped down an old mine shaft, 122.5 m deep, and it reaches the bottom in 5 s. Has the acceleration of the brick due to gravity the value given above these questions?

3. A train travels 40 km at an average speed of 100 km h^{-1} and then it waits for 6 minutes at a station. It then travels for half an hour at 60 km h^{-1}. Find (a) the total distance travelled, and (b) the average speed for the whole distance.

4. The speedometer of a motor-car reads 60 km h^{-1} at the beginning of a certain journey. After driving for a certain time at this speed the driver then changes to 70 km h^{-1}, and again after some time at this speed the driver changes to 80 km h^{-1}. Is the average speed for the whole journey 70 km h^{-1}? Explain.

5. Find the average speed in m s^{-1} during each race at a school athletic sports if the following results were recorded:

100 metres in 10.4 seconds,   200 metres in 21 seconds,
400 metres in 50 seconds,   800 metres in 2 minutes, and
1 000 metres in 2 minutes 40 seconds.
Are these good times for schoolboys?

6. The Jamaican Olympic Games team at Helsinki in July 1952 ran the 4 × 400 metre relay in 3 minutes 3.9 seconds. What was this average speed in metres per second?

7. A boy dives into a swimming pool from a height of 9.8 m. What is the time he takes in falling and what is his velocity on reaching the water?

8. A juggler throws a club up into the air to a height of 4.9 m. How long is it in the air until he catches it?

9. A cricketer throws a ball up 14.7 m. With what vertical velocity must the ball leave his hand?

10. Which produces the most noticeable effect on the human body—a high speed or a high acceleration? Illustrate your answer by referring to travel in a motor-car or an aircraft.

11. Would it be a good idea to add figures indicating distances needed for stopping with good brakes on a dry road on to motor-car speedometers by the side of the figures indicating speeds? If so, would this be possible?

12. Explain all the conditions necessary to fulfil the expression 'What goes up, must come down.'

8. Forces

What causes the acceleration and the retardation of a body? Consider the cases of the falling stone, the ball being thrown upwards into the air and the train during its journey as described by the graphs in Figures 7.2, 7.3, and 7.4 in the last chapter. Forces due to the attraction of gravity act constantly and continuously on the falling stone and on the ball. These create constant accelerations in these bodies. In the case of the moving train variable forces act—the engine varies its force pulling the train and the brakes acting on all the wheels vary their forces—to produce different accelerations and retardations at different times during the journey.

Numerous observations of this nature lead us to the conclusion that a force can give a body an acceleration or retardation and that the value of the acceleration or retardation produced is proportional to the force.

This conclusion can be expressed in this way:

Acceleration ∝ Force

It is common knowledge that a body with a larger *mass* is harder to accelerate than a body with a smaller mass. It is easier to push a stationary bicycle until it reaches walking speed than to do the same thing with a motor-car. The force needed to stop a person who comes running at you is less if he is a small person than if he is a large person. Further observations lead us to the conclusion that the acceleration or retardation given to a body by a certain force is inversely proportional to the mass of the body, and this is expressed in this way:

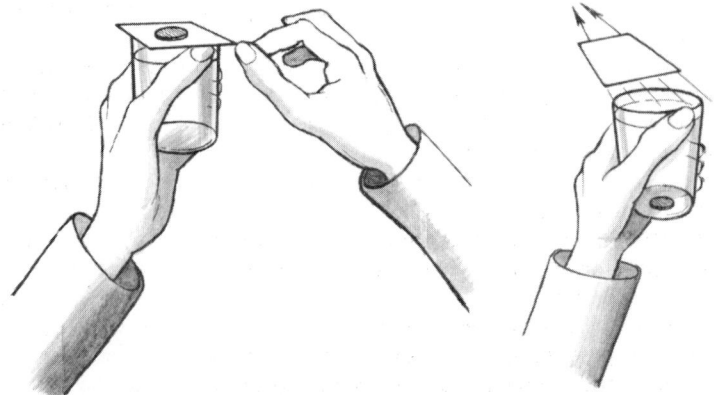

Figure 8.1. On the top of this tumbler of water is placed a postcard and on the card is a penny. Flick away the card horizontally with a quick snap of the finger and the penny will fall into the tumbler. Why will this not happen if the card is moved away slowly?

$$\text{Acceleration} \propto \frac{1}{\text{Mass}}$$

Mass is a word that needs to be defined—it is a property of the body that is accelerated. It is determined only by the quantity of matter in the body and this is sometimes expressed by saying that a body possesses a certain *inertia*. This word inertia describes a fundamental property of a body, and mass tells the amount of inertia in that body. Inertia is a qualitative word and mass is a quantitative word.

The inertia property of a body is well known. It is observed when a passenger in a car is thrown forward on the application of the brakes, and when he is 'left behind' and so is pushed in the back as the car accelerates. The inertia of a glass of water prevents it being swept off the table if the table-cloth underneath it is given a quick sharp pull.

When the earth attracts a body there is a particular force acting on the body that is called the *force of attraction of gravity*, or more simply the *weight* of the body. This particular force gives to the mass of the body the acceleration called g. As we noted earlier g varies slightly according to the position of the body on the surface of the earth. It is difficult to measure g directly as the time taken by a body in falling easily measured distances is very short and the air retards its descent.

Let us examine the two relationships above concerning acceleration. If we combine them we get: force $\propto$ mass $\times$ acceleration, so that the force of attraction of gravity on a mass of 1 kilogramme gives to a mass of 1 kilogramme an acceleration of g metres per second per second. We take as our unit of force a *newton* which is much smaller because it can only give to the mass of 1 kilogramme an acceleration of 1 metre per second per second. It is g times smaller. Thus we can put the value of the newton in this way: a force of g newtons is the same force as the gravitational force on a mass of 1 kilogramme. *The force of gravity acting on an apple is about one newton.*

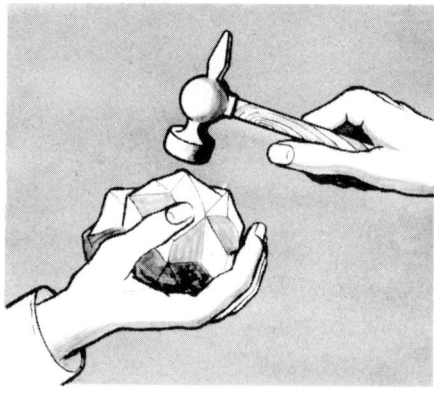

Figure 8.2. Why can this stone be broken by severe blows from the hammer without it hurting the hand of the person who holds the stone?

E

Pilots of high-speed aircraft flying round sharp turns experience forces that give them accelerations of the order of several g. Astronauts have to withstand not only accelerations and retardations of the order of many times g but also, in their weightless conditions, no g at all.

Thus mass and weight are different. *Mass* is the amount of matter in a body, or a measure of its inertia, and it is measured in kilogrammes. *Weight* is the force or pull of gravity on the body and the units in which it is measured are newtons.

When we go to buy some apples we ask for so many kilogrammes of apples and the shopkeeper may measure this quantity by measuring with a spring balance the force of attraction of gravity on the apples. This is the simplest way he has of measuring a quantity of apples. The shopkeeper sells us the apples, that have, for example, a weight of so many newtons, but he says simply and briefly 'Here are six kilogrammes of apples.'

Strictly speaking we should ask for so many newtons of apples and then the shopkeeper can accurately use a spring balance or similar scales using a spring as a control to 'weight' the apples. Then the scales would be calibrated in newtons as some spring balances are now in some schools. But we are more interested in the amount or quantity of what we buy rather than the weight so it would be much more sensible to ask for a mass of six kilogrammes of apples and let him measure the apples by an apparatus which really measures the forces of attraction of gravity on the apples.

Of course it would be better to label, for example, packets of tea or butter or chocolate as 'net mass 500 g' or 'net mass 1 kg' rather than 'net weight' as we do at present.

It is interesting to note that the mass of a body is always constant whereas the weight of a body varies with its position—at sea level the weight is slightly more than on the summit of a mountain and it is very much more on earth than on the surface of the moon.

Let us consider again the relationship between force and acceleration:

$$\text{Force} \propto \text{Acceleration},$$

$$\text{or } f \propto a.$$

and the relationship between mass and acceleration:

or
$$a \propto \frac{1}{m}.$$

Combining these two relationships as stated earlier we have mathematically:

$$a \propto F$$

and
$$a \propto \frac{1}{m}$$

so that
$$a \propto \frac{F}{m}$$

or $\quad\quad\quad\quad\quad\quad\quad F \propto m \times a$

or $\quad\quad\quad\quad\quad\quad\quad F = kma.$

We have so designed the units that k the constant in this equation becomes unity.

$\therefore$ Force (N) $=$ Mass (kg) $\times$ Acceleration (m s^{-2})

Example 1. A body of mass 10 kg is raised vertically on a string with a force of 150 N. What is the acceleration produced in the body?

The force producing the upwards acceleration is the resultant of two forces —one the force of 150 N upwards and the other the force of attraction of gravity on the mass of 10 kg downwards namely 9.8×10 N.

$$\text{This force is} = 150 \text{ N} - 9.8 \times 10 \text{ N}$$

$$= 52 \text{ N}$$

$$\text{But force (N)} = \text{mass (kg)} \times \text{acceleration (m s}^{-2})$$

$$\text{or } 52 = 10 \times \text{acceleration}$$

$$\therefore a = 5.2 \text{ m s}^{-2},$$

$\therefore$ the acceleration of the body upwards is 5.2 metres per second per second.

Example 2. A train of mass 100 000 kg has an engine that hauls it with a steady force of 400 000 N. How long will it take before the train reaches a speed of 50 km h^{-1}?

Assume g to be 10 m s^{-2}.

$F = 400\,000$ N, $m = 100\,000$ kg, find a.

Substitute in the equation

$$F = ma$$

$$\therefore a = \frac{400\,000}{100\,000} \text{ m s}^{-2}$$

$$= 4 \text{ m s}^{-2}.$$

Using the same abbreviations as in the last chapter the speed of the train is given by:

$$v = u + at.$$

$u = 0$ m s^{-1}, $v = 50$ m s^{-1}, $a = 4$ m s^{-2}, and $t = t$ s

$$\therefore 50 \text{ m s}^{-1} = 0 \text{ m s}^{-1} + 4 \text{ m s}^{-2} \times t,$$

$$\therefore t = 12.5 \text{ s}.$$

$\therefore$ the train reaches a speed of 50 kilometres per hour in 12.5 seconds.

Pendulum. We have said that it is difficult to measure '*g*' directly but Galileo found how to do it when he discovered the principle governing the to-and-fro motion of a pendulum. He found that the time needed for a complete vibration over to one side and back again depends only on the length of the pendulum and the acceleration due to gravity. The ideal pendulum has a thin weightless string and heavy bob of negligible size suspended from a single point. When the bob swings through a small arc the time of a complete vibration (*t*) is given by the equation:

$$t = 2\pi \sqrt{\frac{l}{g}}$$

where *l* is the length from the actual point of support to the centre of gravity of the bob.

Gravity enters into this equation because the bob at either end of the arc is just a little higher than it is at the centre of the swing. The force of attraction of gravity, *g*, pulls the bob downwards and because it is attached to the string makes it swing. Set up a pendulum for yourself and count the time of 30 swings and then repeat this using (*a*) a heavier bob, (*b*) a bob of another material, and (*c*) a bigger arc of swing. Does the time of a complete vibration vary?

Does the acceleration due to gravity of a heavier bob or a bob of another material vary from that of the original bob? Can you think of any connection between these two questions?

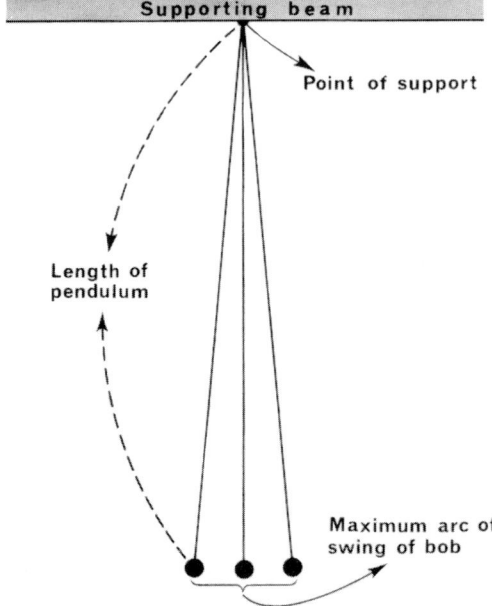

Figure 8.3. A simple pendulum. The point of suspension must be rigid and the arc of swing must be small.

Questions on chapter 8

Assume g to be 10 m s^{-2}.

1. What is meant by the inertia of a body?

2. Why can quite a small stone thrown up by a passing car shatter the windscreen of a car travelling in the opposite direction?

3. Why does a stone thrown across smooth ice on a lake sooner or later stop moving and come to rest? Why do not the planets similarly come to a stop?

4. Explain why when a heavy bomb dropped from a high-flying aircraft falls toward the earth, the earth rises up to meet it.

5. Is it easier to walk towards the front of a train when it is travelling at a high speed or when it is coming to a halt? Give complete reasons for your answer.

6. A man whose weight is 800 N enters a lift. He stands on a weighing machine on the floor of the lift. What will the machine register when (a) the lift is rising steadily at 3 m s^{-1}, (b) the lift is accelerating upwards at 1 m s^{-2}, and (c) the lift is accelerating downwards at 2 m s^{-2}?

7. At the start of a sprint an athlete accelerates at the rate of 2 m s^{-2}. What force does he exert backwards against the starting block if his weight is 600 N?

8. A boy standing in a lift is holding a parcel of mass 8 kg by a string that will break when a force of 100 N is applied. What is the maximum acceleration that the lift can have before the string breaks?

9. A motor-car weighing 11 000 N starts from rest and accelerates steadily until it is travelling at 36 km h^{-1}. If it takes 11 s to attain this speed what is its acceleration? What is the total force exerted by the tyres on the road to produce this acceleration? How far has the car travelled during these 11 s? (Neglect air resistance and other frictional forces.)

10. What retardation will a motor-car weighing 19 800 N have due to its brakes if they apply a total braking force of 4 950 N? How long will these brakes take to bring the car to rest from 72 km h^{-1}? How far will the car travel before stopping? (Neglect air resistance and other frictional forces.)

9. Forces in action

We have considered the effect of *one* force acting on a body and the acceleration or retardation it produces. In this chapter we shall consider the effect of more than one force.

It is usual and also helpful to represent a force by drawing on a flat sheet of paper a straight line in the direction of the force and making its length equivalent on some scale to the strength of the force. All the problems we need to solve can be done with scale drawings of forces in this manner.

If two forces act along the same straight line in the same or in opposite directions they may be represented by a single force whose strength is obtained by adding or subtracting the strengths of the two forces. This single force is called the *resultant* of the two forces.

We can perform an experiment in order to find the resultant of two forces meeting at a point and inclined to one another at an angle. Suspend over two smoothly running pulley wheels a long string on the ends of which are two masses of 4 kg and 3 kg each. Between the pulley wheels tie another string on the end of which is a third mass of 5 kg. Figure 9.3a shows how this is done. These three masses produce three proportional forces in newtons. These three forces are in equilibrium which means that they balance one another and do not move from the position they take up. Mark the directions of the three strings meeting at O on a sheet of paper. On this sheet of paper make a

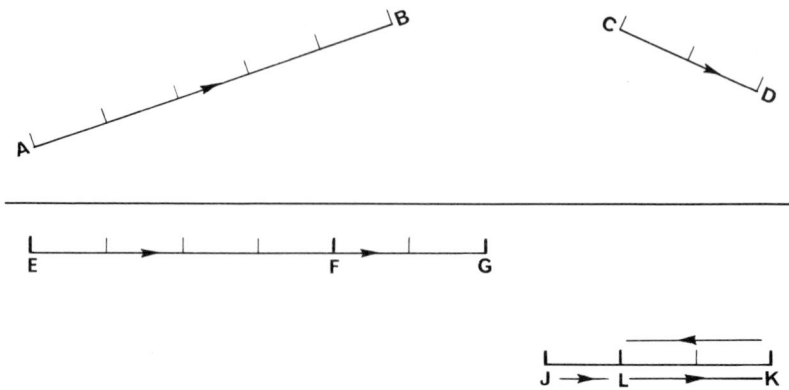

Figure 9.1. These lines represent two forces AB of 500 units and CD of 200 units acting in the direction of the lines from A to B and from C to D respectively.

Figure 9.2. The resultant of the two forces EF of 400 units and FG of 200 units acting along the same straight line and in the same direction is EG of 600 units. The resultant of the two forces JK of 300 units and KL of 200 units acting along the same straight line but in opposite directions is JL of 100 units in the direction J to L.

scale drawing of the three forces (Figure 9.3b), using the directions copied from the strings and the strengths of the forces as read on the spring balances. All the three forces cancel one another because the junction at O does not move. We say that the three forces are in *equilibrium*. The resultant of the two forces OA and OB must therefore be equal in strength and opposite in direction to the force OC. We now observe that if we draw OC in the opposite direction as OD (Figure 9.3c) the geometrical figure OADB is a parallelogram.

Hence in order to find the resultant of two forces we complete the parallelogram formed by the two forces and draw in the diagonal from the point where the two forces meet. This diagonal represents the resultant force that can replace the two original forces.

If there are more than two forces meeting at a point the resultant can be found by drawing in pairs forces and resultants until one final resultant is left.

It frequently happens that a single force acts on a body in one direction and produces results as though it acted like two or more forces in other directions. These are known as *component* forces. See Figure 9.4.

We can use the parallelogram to find the component forces of a single force in the same way that we use a parallelogram to find the resultant of two component forces. The single original force is made the diagonal of a parallelogram and the component forces are represented by the two adjacent sides about the diagonal.

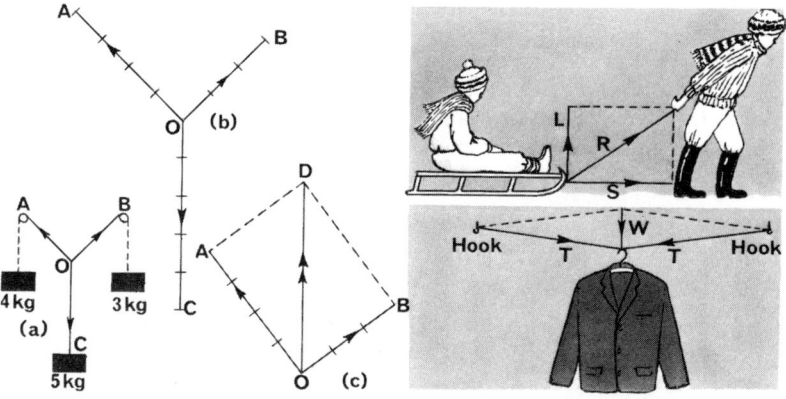

Figure 9.3. An experiment to find the resultant of two forces acting together at a point but in different directions. The resultant of forces OA and OB is OD because they are balanced by the equal and opposite force OC. The forces OA, OB and OC are the forces due to gravity acting on the masses 4 kg, 3 kg and 5 kg respectively.

Figure 9.4. The components of a force obtained by the scale drawing of parallelograms. (a) The boy pulling the loaded sleigh over the snow exerts a force R that creates one component force S along the surface of the snow causing the sleigh to slide along, and another component force L tending to lift the front edge of the sleigh off the snow. (b) A wet coat hanging to dry on a taut clothes line creates two equal and very large forces tending to pull out the hooks at either end. The weight of the wet coat is W and the force tending to pull out a hook is T.

Moments. A force can cause a body to turn about a pivot or a *fulcrum*. The product of the force (in N) and the perpendicular distance (in m) between the line of action of the force and the fulcrum is called the *moment* of the force about the turning point.

Many moments can be applied to a body—some tending to turn it in one way and some in the other. When the sum of all the clockwise moments equals the sum of all the anticlockwise moments the body does not turn at all and is said to be in *equilibrium*.

The wheelbarrow in Figure 9.5 is lifted by a vertical force, called the *effort*, and this is applied to the handles. The weight of the wheelbarrow and its contents, called *load*, acts vertically downwards as shown by the arrow. These two forces produce moments about the axle of the wheel, called the fulcrum, and when the barrow is steady these two moments are *equal in magnitude* but *opposite in direction*. This is expressed by the equation:

CLOCKWISE MOMENTS = ANTICLOCKWISE MOMENTS,

or,   LOAD × DISTANCE OF LOAD = EFFORT × DISTANCE OF EFFORT,
   from fulcrum   from fulcrum

Given the data shown in Fig. 9.5 therefore:

$$250 \text{ N} \times 1 \text{ m} = \text{Effort} \times 2 \text{ m},$$

∴ Effort needed to lift the

wheelbarrow = 125 N.

A lever is a bar or rod that can swivel freely about a fulcrum. The simple beam balance is a lever in which the load arm is equal in length to the effort arm. When the beam is balanced the moments on both sides of the pivot are equal in magnitude and opposite in direction to one another. Therefore because the

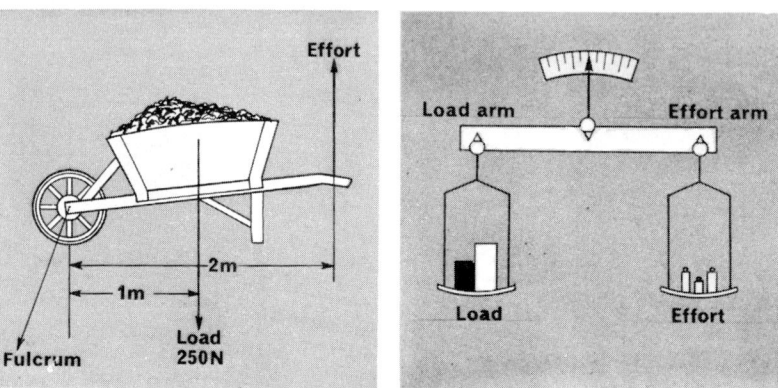

Figure 9.5. Two moments act about the axle of the wheel of the wheelbarrow. The load produces a clockwise moment and the effort an anticlockwise moment.

Figure 9.6. A simple beam balance. The chemical balance is a very accurate instrument built in a similar way. Strong beam balances are also made for use with heavy loads.

two arms are equal in length the load must equal the effort in magnitude. The level balance or beam balance provides a simple way of comparing the weights of two bodies.

There are many other ways of using levers—most of them have unequal lengths of arm for the load and effort so that a greater or smaller load can be moved compared with the effort applied. (See Figure 9.7.)

Parallel forces. Some of the examples we have dealt with so far concerned forces that are parallel (Figure 9.6) and some are not (Figure 9.4).

Bodies are often supported or carried by two parallel forces; for example, a bridge supported by two piers, or a pole carried by two persons one at each end. The resultant force of these two parallel forces is the weight of the body supported. The sum of the strengths of the two parallel forces is the strength of the resultant and they all three act in the same direction. The actual line along which the resultant acts is such that the moment of one of the parallel forces is always opposite to the moment of the other parallel force about the resultant. The two examples drawn in Figure 9.8 will make this clear.

Couples. Two parallel, equal, and opposite forces acting on a body form a *couple* and these produce a moment that causes the body to turn. The body will continue to turn if the couple persists. Two couples of the same magnitude and acting in the opposite sense will balance one another and keep the body in equilibrium.

There are three examples of bodies on which couples act shown in Figure 9.9. A couple is exerted by the box spanner and tommy bar on the bolt (*a*) causing it to turn until an equal and opposing couple is created by the frictional forces between the bolt and the frame. Then the bolt cannot be tightened further. A pilot of an aircraft in steady flight (*b*) balances the clockwise couple of drag and thrust against the anticlockwise couple of lift and weight. Care is taken in the design of the aircraft to enable the pilot to achieve this balance easily by the use of small trimmers. A yacht (*c*) is sailed so that the two couples balance.

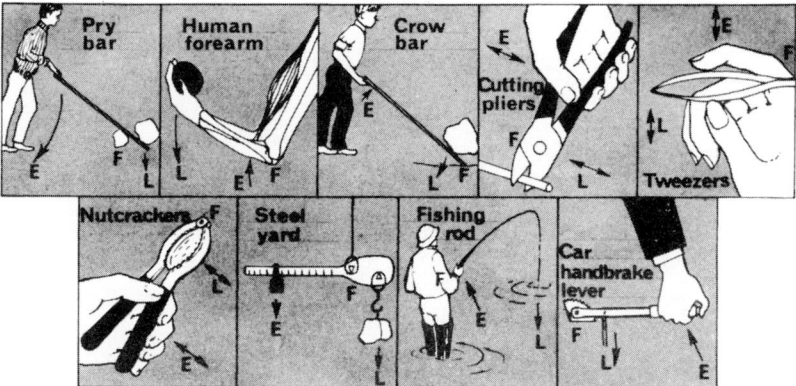

Figure 9.7. Various types of levers. The letters F, L, and E are marked below the fulcrum, load, and effort in each case. The arrows show the directions of the forces of the load and effort on the lever when it is balanced.

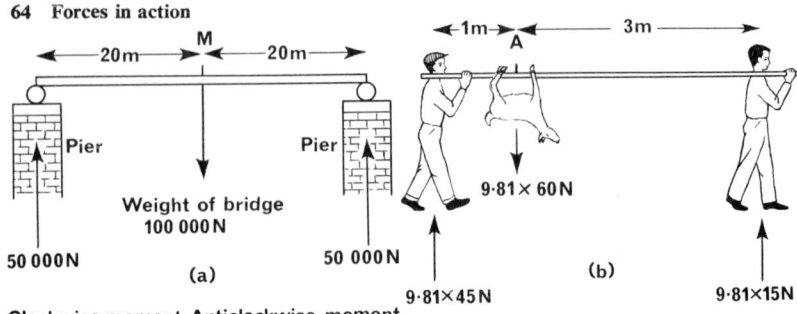

Clockwise moment   Anticlockwise moment
        about M   about M
50 000 N × 20 m   50 000 N × 20 m
=1 000 000 Nm   =1 000 000 Nm

Clockwise moment. Anticlockwise moment
        about A   about A
9·81×45N ×1m   9·81×15N ×3 m
= 9·81 ×45Nm   = 9·81×45Nm

Figure 9.8 (*a*) A bridge 40 m long supported on two piers. (*b*) A man and a boy carrying a heavy animal on a light pole 4 m long. The moments acting around each weight counterbalance one another. The piers of the bridge carry equal weights, but the man carries three times the weight the boy carries. What is the position of the weight in each case?

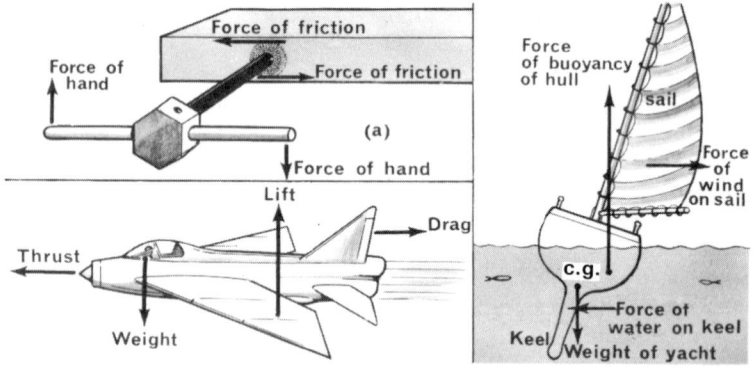

Figure 9.9. Three examples of balanced couples acting on bodies to keep them in equilibrium.

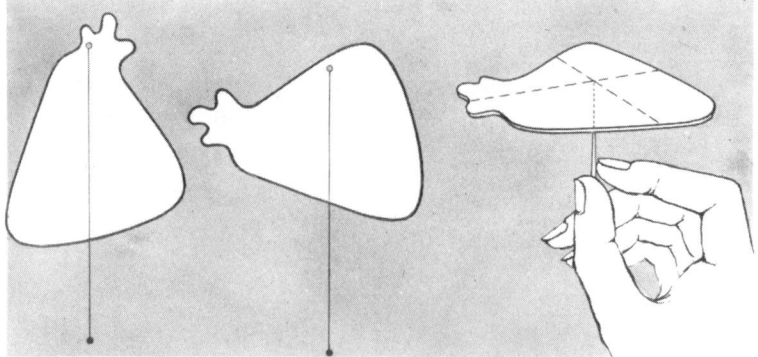

Figure 9.10. These diagrams show how to find the centre of gravity of an irregular shaped flat body. Hang it up from any point by the side of a plumb line and mark a line across its surface along the plumb line. Repeat this several times from different points of suspension. Where all the lines intersect is the centre of gravity of the flat body. The body will balance perfectly on a fine point held under this centre of gravity.

As the force of the wind on the sail varies so the angle the yacht heels over varies and moves the position of the force of buoyancy.

Centre of Gravity. A series of parallel forces act on a body due to the attraction of gravity. In fact, on every particle there is a force directed to the centre of the earth. The sum of all these forces is the weight of the body. All these many small forces can be replaced by its weight acting as a single force through a point known as the *centre of gravity* (c.g.), or sometimes as the centre of weight.

The position of the centre of gravity of any body whether solid or flat can be determined because it is always vertically below its point of suspension if it is allowed to hang freely—the further it is below the steadier the body becomes. If it is balanced on a point the centre of gravity is always vertically above the point of support but it is extremely difficult to put it there and to keep it there.

In the case of a solid body it is almost always impossible to touch the centre of gravity because it is inside the body but its position can be estimated from outside.

Stability. The position of the centre of gravity in a body determines the stability of that body. If a vertical line through the centre of gravity falls well within the base of the body it is in *stable equilibrium* and will, if disturbed slightly, fall back on its base. If, however, this line falls on the edge or outside the base a slight disturbance will cause it to fall over, and it is then said to be in *unstable equilibrium.*

The loading of freight and passengers on vehicles should be done with forethought, for if the centre of gravity of the load is too high there is a possibility that in travelling over a rough piece of road a state of unstable equilibrium will be reached.

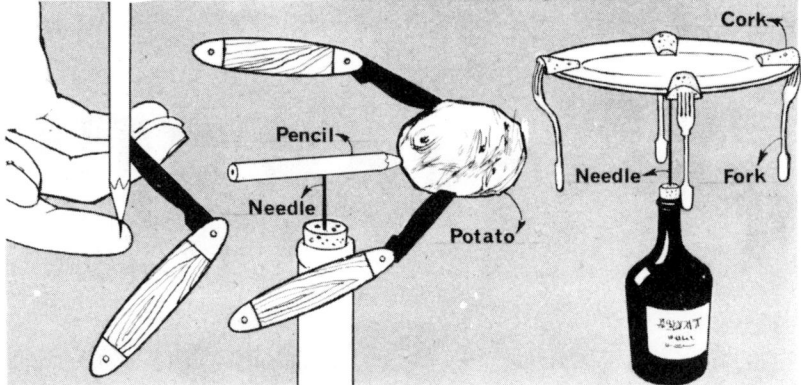

Figure 9.11. Some simple balancing tricks worth trying. In each case estimate where the centre of gravity is to be found.

Figure 9.12. Stable and unstable equilibrium. The boy on the left sits quietly on all four legs of his chair and the vertical through his centre of gravity is easily within the base. The other boy has wriggled so much that the vertical line through his centre of gravity has fallen outside the base and over he will go.

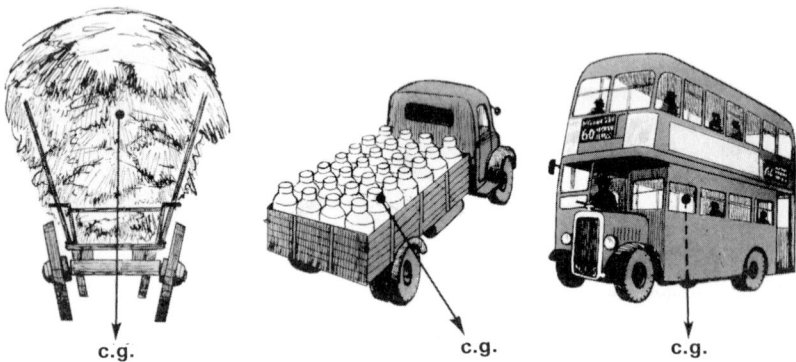

Figure 9.13. How far will each of these vehicles tip over before reaching the state of unstable equilibrium? Why should the passengers in the bus fill the seats downstairs first before occupying seats upstairs?

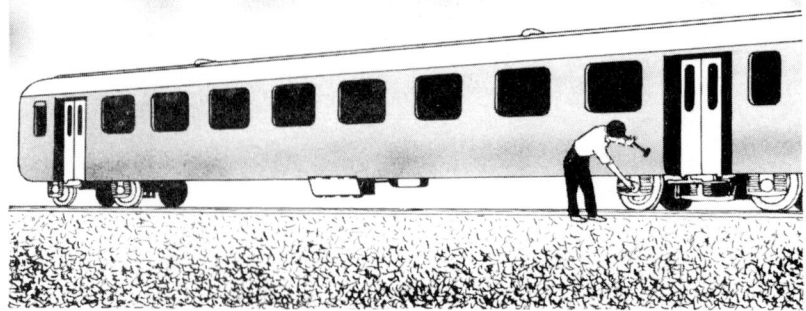

Figure 9.14. Axle bearings are tested frequently during the journey of a railway train. 'Hot boxes', due to insufficient lubrication and thus excessive friction, often cause accidents and delay.

Friction. The force that opposes the motion of one surface moving over another with which it is in contact is called the *force of friction*. Its magnitude depends on the materials of which the two surfaces are made, as well as on the force pressing them together. Some of the energy put into machines is transformed into heat energy because of the friction between moving parts. The heat may cause serious damage in addition to being wasteful.

We cannot get rid of friction entirely but we can reduce it considerably by a suitable choice of surfaces and by lubrication.

To start a body moving over another with which it is in contact, a force at least as great as the *starting friction* has to be applied; this force of starting friction is greater than the force called *sliding friction* which is needed merely to keep the body moving. A very considerably smaller force called *rolling friction* is sufficient to keep one body moving against another if the two surfaces facing one another are hard and there are hard rollers or balls between them.

It is important that the metal surfaces of *roller or ball bearings* that come in contact should be really hard. If one of the surfaces is not hard then the rolling friction might well be more than the sliding friction; it is for this reason that aircraft landing on soft snow fit skis in place of wheels. (Figure 9.16.)

Oil or grease is placed between the moving surfaces to reduce friction still further, for surfaces slide on grease more easily than on one another.

Friction comes to our aid in some circumstances and we try to increase it, not to reduce it. Without friction we would be unable to walk about for we would slip and slide in every direction. It is friction that enables nails to hold two pieces of wood together. The force of friction between wheels and brake blocks slows the movement of a bicycle. Belt drives on machinery do not slip because of friction. In some bad weather conditions the wheels of locomotives do not grip the railway lines unless the engineer throws fine sand between the wheels and the lines to increase the frictional force.

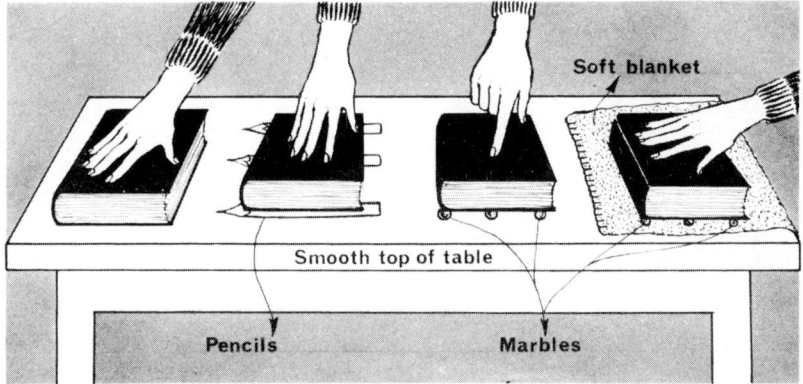

Figure 9.15. This is an experiment to demonstrate how the force of friction changes with the surfaces and the bearings used. The difference can be detected easily with the hand.

Questions on chapter 9

1. Two forces, of 100 N and 200 N respectively, are inclined to each other at an angle of 60°. Find by scale drawing the magnitude of the resultant and the angle between it and the force of 200 N.

2. A picture weighing 200 N is hung in the usual way from a single hook. If the angle made by the wires at the hook is 60°, what is the tension in the wires?

3. A yacht is 'tacking' against the wind. The wind produces a force at right-angles to the sail. This force can be represented by two component forces, one pushing the yacht forward and the other driving it sideways. Draw a diagram showing the yacht, its sail, the direction of the wind, the force on the sail and its two component forces. How is the action of the component force driving it sideways balanced?

4. A see-saw is 6 m long and is pivoted in the middle. A body of mass 60 kg is suspended from one end by a string. How far from the middle must a 75 kg mass be suspended in order that a balance may be obtained?

5. What is meant by the moment of a force about a point?

How is a moment balanced? Use a simple lever to illustrate your answer.

A metre rule has a hole drilled in the centre at the 1 cm mark. A knitting needle is put through the hole and fixed to a support so that the ruler can swing in a vertical plane. Two horizontal forces are applied to the rule, one of 100 N at the 41 cm mark, and the other of 50 N at the 61 cm mark. They act in such a way as to swing the ruler from the vertical position, both acting in the same direction. What is the single horizontal force acting at the 81 cm mark which will bring the ruler back to the vertical?

6. Explain how you could lift a large mass by means of an iron bar used as a lever. (Two possible answers.)

A uniform rod, 100 cm long, balances on a knife-edge placed at a distance of 30 cm from one end when a mass of 0.75 kg is suspended from that end. Calculate the mass of the rod.

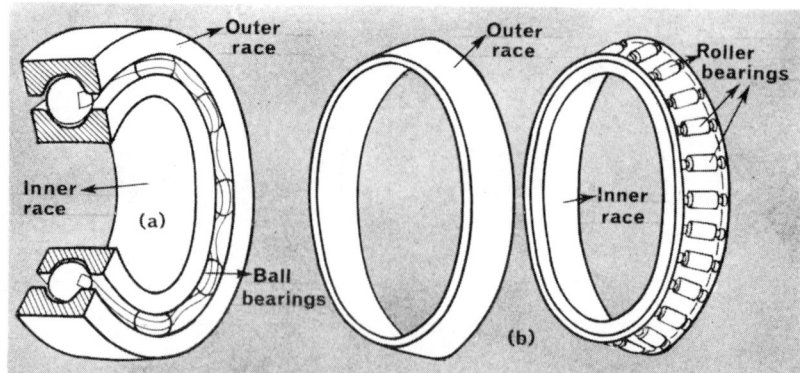

Figure 9.16. Bearings. (*a*) A cut-away section of ball bearings showing how the steel balls are held in position between the inner and outer races. (*b*) Roller bearings mounted in a sloping position so that fine adjustments can be made to the two races in order to eliminate any 'play'.

7. Describe how you would find the centre of gravity of a flat piece of wood of irregular shape.

A plank 6 m long, is balanced at its centre to make a see-saw. A boy of mass 42 kg sits 2 m from the centre, and another boy of mass 30 kg sits 1 m from the centre, both on the same side. Calculate how far on the other side a man of mass 57 kg must sit to balance the two boys.

8. A truck of sugar beet weighing 30 000 N is passing over a bridge 200 m in length. If the truck is 40 m from one end of the bridge, how much of the weight is supported by each pier? Make a diagram to show the direction and the strengths of the forces in newtons.

9. A block of wood is pulled along a flat horizontal surface by a force parallel to the surface. Draw a diagram showing the force pulling the block, the weight of the block, the force of friction, and the reaction of the bench.

10. Why do racing yachts with large sails have very heavy keels?

11. Why is a loaded double-decker bus more unstable than the same bus without any passengers upstairs?

12. Draw diagrams that will explain the reasons for the following:

(a) when a small quantity of lead shot is placed in a test-tube, the test-tube floats upright when placed in still water;

(b) the hands of a large clock are extended backward beyond the central spindle;

(c) most chairs have legs that slope outwards;

(d) an archer stands with his legs well apart when firing his arrow; and

(e) drinking glasses often have wide bases.

10. Some properties of solids

Dimensional changes due to forces. The study of the properties of solid materials is of great importance to designers and builders. They need to know what external forces wire cables can withstand before they break, what is the best material for the springs of vehicles, and what amount of bending of the metal can take place at the junction of the wing and the fuselage of an aircraft before it collapses. The answers to these questions, and to many others, requiring a knowledge of the strengths of materials, are largely determined by the forces between the molecules of the materials themselves.

Some metals, notably steel, resist very strongly any forces that try to pull their molecules apart. Other metals, like copper, can be stretched easily and are said to be *ductile*. Others, like gold and lead, can be beaten by hammers or rolled into thin sheets—gold leaf is a very thin sheet of gold; these metals are *malleable*.

Another property of some metals is that although they suffer some distortion (*strain*) due to external forces (*stress*) they can recover from this distortion immediately the forces are removed. This property is known as *elasticity*. Steel is a good example of a material that is almost perfectly elastic for it

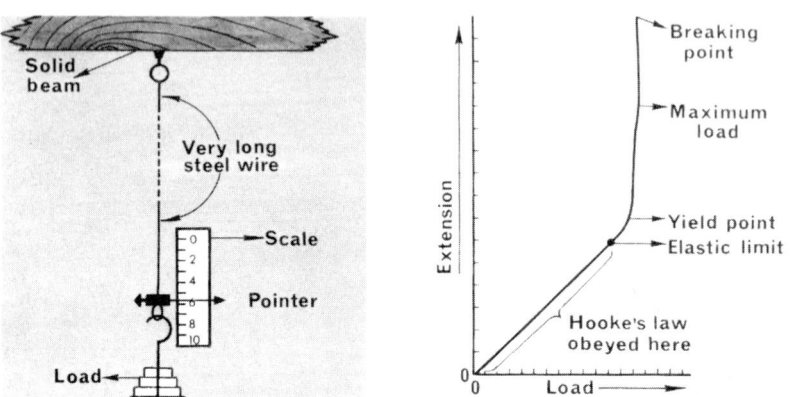

Figure 10.1. How the elongation of a long steel wire under load is observed. The long steel wire is attached to the solid beam above and a load is hooked on to its lower end. The elongation is observed by the movement of the pointer over the scale. This is repeated with different loads.

Figure 10.2. The graph shows the load applied and the extension produced in a steel wire. The steel wire is perfectly elastic until the elastic limit is reached, and after that the wire stretches easily for a small increase in load until a position of maximum load is reached. The wire then becomes weak, supports less load, and very quickly breaks altogether.

70

requires a large force to deform it but once the force is removed the deformation disappears. Plastic materials have extremely little elasticity.

It is possible to observe the elongation that takes place when a wire of some elastic material is stretched. The results of a series of such observations made with a steel wire using the apparatus as shown in Figure 10.1 are best examined after being plotted on a graph like the one drawn in Figure 10.2.

The important feature of this experiment, shown clearly in the graph, is the straight line section as far as the elastic limit. Over this section the elongation is proportional to the load. This property was found by an Englishman, Robert Hooke, three centuries ago. It is now known as *Hooke's Law*. It is usually stated simply in these words: 'Within the limit of elasticity the elongation is proportional to the load.' Or expressed by an equation it is:

$$\frac{\text{LOAD (or Stress)}}{\text{ELONGATION (or Strain)}} = \text{CONSTANT.}$$

Example. The pointer attached to the long steel wire in Figure 10.1 reaches to the scale mark 3 when a load of 15 N is hung on it, and to the scale mark 5 when the load is 95 N. What load will bring the pointer to the 4 scale mark, and what will be the reading of the pointer on the scale when a load of 75 N is applied?

These two pairs of readings can be plotted on a graph and a straight line drawn through them; then the answers can be read off easily. Figure 10.3 shows how this is done.

From the 1 mark on the elongation scale which occurs when the pointer is at scale mark reading 4 draw a horizontal dotted line to cut the straight line, and from there draw a dotted line vertically. This vertical cuts at 55 N. Hence the load required to lower the pointer to the scale mark 4 is 55 N. By drawing similar dotted lines it is observed that the scale mark reached by the pointer for a load of 75 N is 4.5.

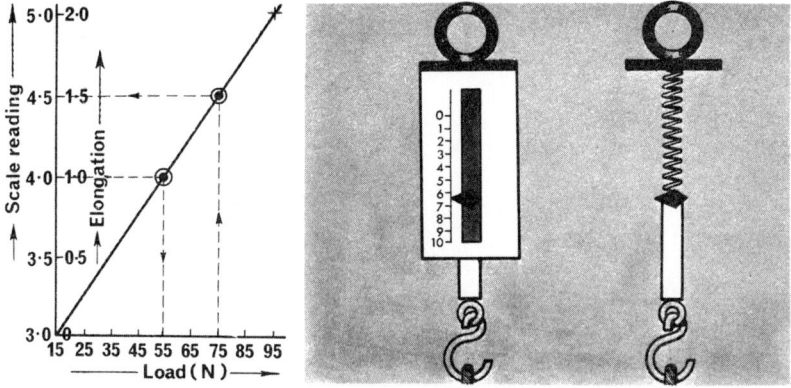

Figure 10.3. A graph of the load and elongation of a perfectly elastic steel wire.

Figure 10.4. A complete spring balance and on the right its spring and pointer removed. The spring when elongated to the end of the scale is well within the elastic limit of the spring so that it always returns to a correct zero reading.

F

The spring balance is a simple piece of apparatus that enables us to find out how strongly the force of attraction of gravity is pulling a body. The body is simply suspended on the spring by means of a hook and the elongation of the spring is indicated by a small pointer attached to it. The scale on which the pointer moves is so made that it tells us directly the weight of the body. The beam balance enables us to compare the weights of two bodies, but the spring balance compares the weight of a body with the elongation of a spring. In order to show the slight difference in the weights of a body at different parts of the earth a delicate spring balance must be used and not a beam balance, for the latter would be acted on by two weights both of which would have experienced the same change.

Questions on chapter 10

1. State Hooke's law. Draw a labelled diagram of a spring balance in use.
In using a spring balance, the scale readings are found to be (a) 5.5 cm for a load of 100 N, and (b) 9.0 cm for a load of 800 N. What will be the load when the scale reads 7.5 cm?

2. The vertical steel cables used to suspend a bridge stretch 1 cm when a motor-car passes over the centre of the bridge, and 5 cm when a loaded lorry does the same thing. Compare the weights of the car and the lorry. What assumptions must be made in working out this problem?

3. A force of 100 N causes an iron wire to stretch 1 cm. What is the extension produced when (a) three such wires attached end to end, and (b) the three wires clamped together to form a cable three times the cross-sectional area but the original length, are stretched by the same force?

11. Some properties of liquids

Pressures in liquids. We have already studied how forces applied in machines have been transmitted by beams, rods, cables, and ropes. There are machines in which the forces are transmitted by liquids. A force pushing a liquid cannot be applied at a single point—it has to be spread over an area, usually by means of a piston in a cylinder. The pushing force is then often called a *thrust*, and it creates a *pressure* in the liquid measured by the thrust acting on a unit area of the piston, or

$$\text{PRESSURE} = \frac{\text{THRUST}}{\text{AREA}}$$

Pressure is therefore measured in N m^{-2}.

A liquid that is confined in a sealed container can be subjected to a pressure and that *pressure is transmitted* equally in all directions at any part of the container. This property is made use of in many practical applications.

The force or thrust F_1, shown in Figure 11.2, acting on the area A_1 produces a pressure P in the liquid. This pressure is transmitted through the liquid and produces a larger force or thrust F_2 on the larger area A_2. The force or thrust is directly proportional to the area.

It is important that the brakes on the four wheels of a motor-car should be operating evenly at all times. This is achieved in the following way. The foot depresses the brake pedal and this causes a piston to press upon the fluid in the master brake cylinder. Examine Figure 11.4. By means of the fluid in the small

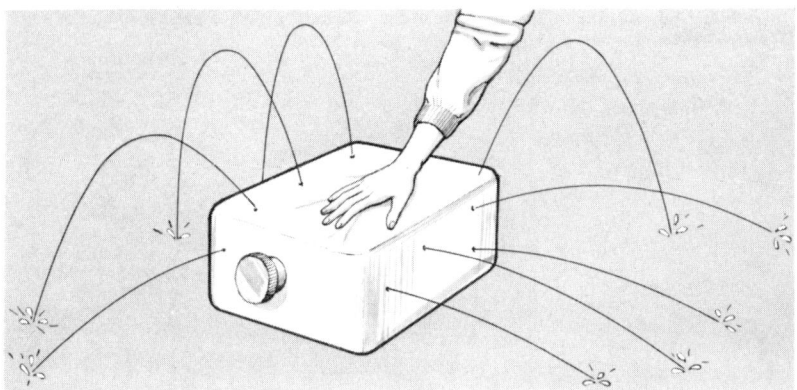

Figure 11.1. A polythene water carrier is filled with water and it is punctured in many places with a fine pin. When a person presses down hard with his hand the water squirts out equally in all directions but when he does not press the water does not leak out of the holes.

73

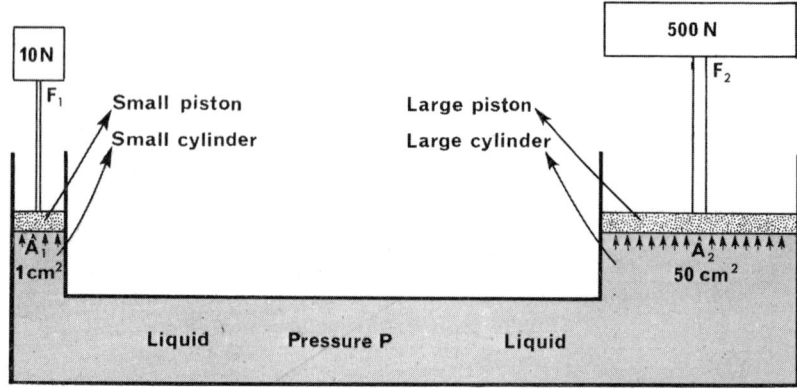

Figure 11.2. How a hydraulic press works.

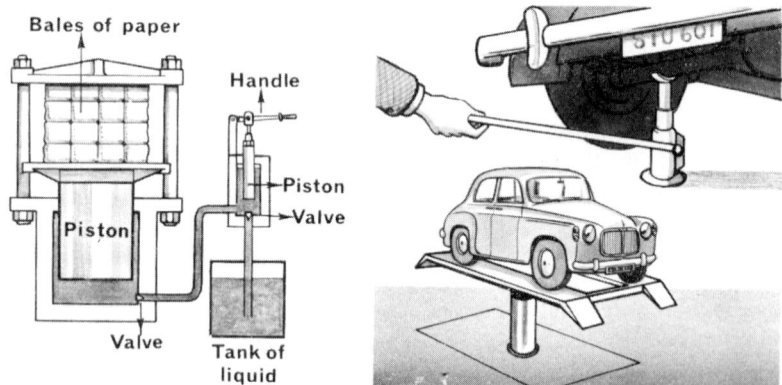

Figure 11.3. Applications of the principle of the hydraulic press—a press for bales of waste paper—a car hand jack—a car hoist.

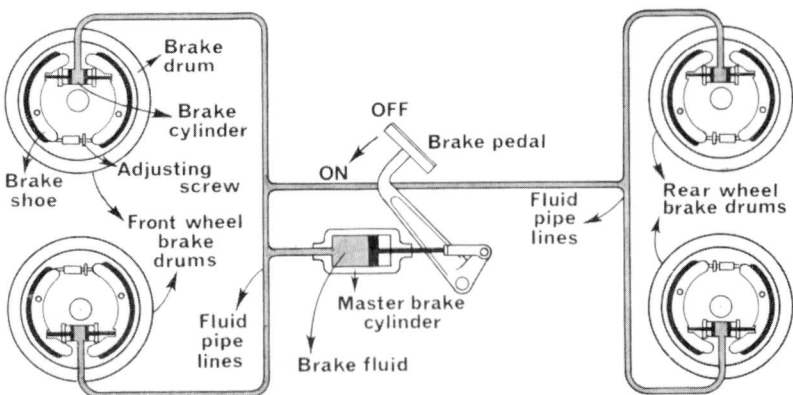

Figure 11.4. The hydraulic braking system of a motor-car.

pipes the pressure exerted by the piston is transmitted to each of the four wheels. In each wheel the pipe leads to a small cylinder fixed inside the brake drum. As the pistons in these small cylinders are forced outwards by the pressure of the fluid they press the brake shoes forcibly and equally against each of the four brake drums. The friction thus produced equally on each pair of wheels stops the car without making it turn or skid.

We have discussed the transmission of pressure from outside through a liquid, but liquids by themselves, because they have weight, exert a pressure on the sides and the base of the vessels that contain them. Fishermen wearing long wading boots feel sideways pressures on their legs when standing in deep water. These pressures are greater the deeper the water.

The *pressure at any depth in a liquid* contained in a vessel must be the same in any direction—sideways, upwards, and downwards—otherwise the liquid would move being pushed on by the greater pressure, and no such movement is observed. The pressure evidently depends only on the depth and not upon the shape or size of the liquid above because we can see that liquids in vessels of various shapes remain motionless. Hence a liquid at rest has a surface that is horizontal, and it does not matter if this surface is in several separate sections provided that all the sections are connected to one another.

Note the *height of the water tower* in Figure 11.7. Water is pumped up there from the storage reservoir after being purified in the filter beds. The water from the tower flows down to the water mains and then rises into the houses and the hydrants as it tries to rise and reach again the same level as it has in the tower. Because the tap in the house prevents the water from rising to its level the water is under pressure at the tap. In which house, A or B, is the water pressure at the tap greater?

Buoyancy in liquids. A person walking across a pebbly beach into the sea to bathe finds that before he enters the water the stones hurt his feet, but that the further he enters the water the less the stones hurt, until finally he begins to float and he then feels the stones no longer.

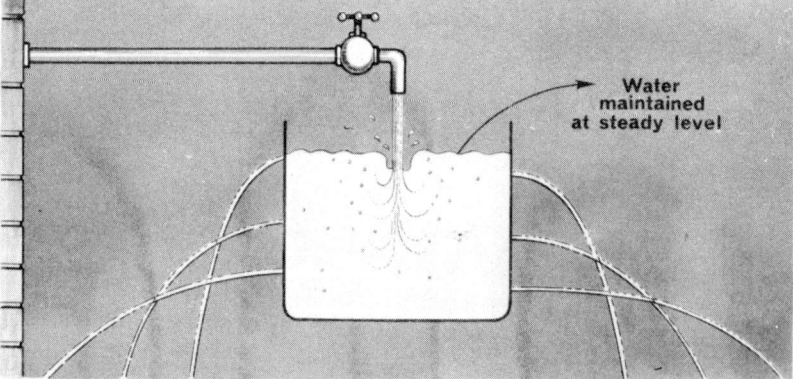

Water
maintained
at steady level

Figure 11.5. Does this experiment show that the pressure at a point below the surface of a liquid varies with the vertical depth?

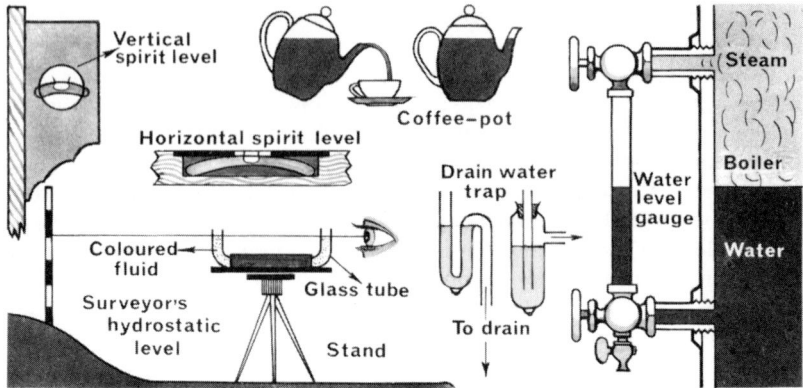

Figure 11.6. Some of the useful applications of the fact that liquids in connecting vessels have their surfaces at the same level.

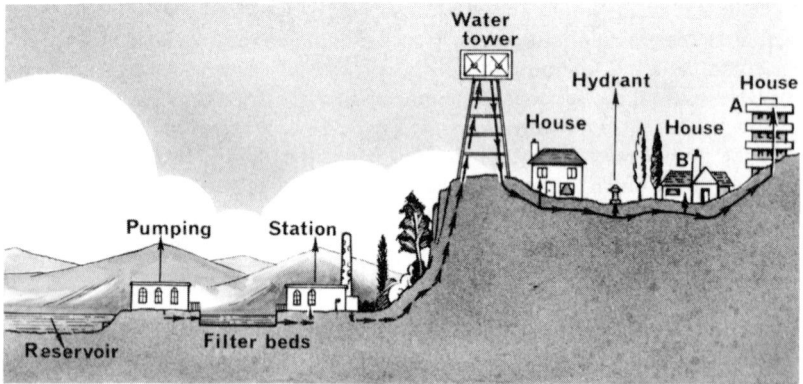

Figure 11.7. A section of a town water supply. The pumped water is shown by the dotted line with arrows and the water that falls under gravity is shown as a continuous line with arrows.

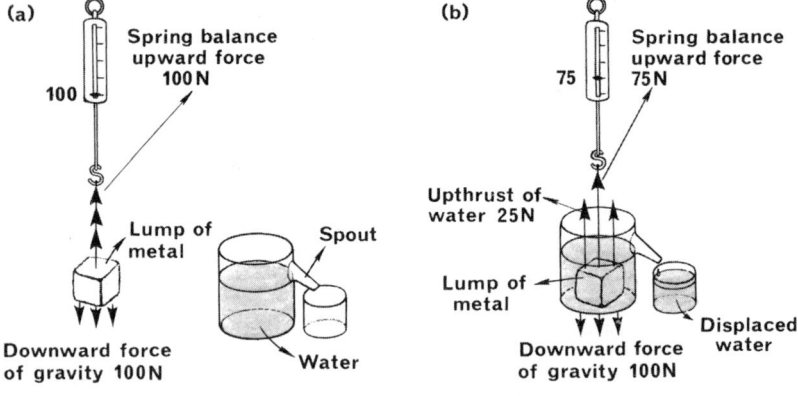

Figure 11.8. An experiment to illustrate the principle of Archimedes. The lump of metal is weighed (*a*) in air, and (*b*) in water.

The explanation of this experience is that the sea water provides a thrust upwards on his body and this upthrust increases as more of his body is below the water surface.

By weighing objects in various liquids and noting in each case the upthrust and the weight of the liquid displaced by the object, a principle first attributed to the Greek philosopher Archimedes can be observed.

Look at Figure 11.8. The lump of metal weighs 100 N in air according to the pull on the spring balance. The can on the left is filled with water up to the spout and the overflow can is empty. The metal is then completely immersed in the can as shown in (b), the water overflows into the small can, and the upthrust reduces the pull on the spring balance. The upthrust is the difference between the two pulls indicated on the spring balance, namely (100 – 75) N or 25 N. The water that overflowed into the small can is also found to weigh 25 N. This result expressed in general terms shows that the upthrust on a body completely immersed in a liquid is equal to the weight of the liquid that the body displaces.

Flotation. When the upthrust on a body partially immersed in a liquid equals the weight of the body then the body floats. In other words, a floating body sinks to such a depth that the weight of the liquid it displaces then equals the weight of the body itself.

A solid lump of iron placed in water will sink because the water it displaces does not weigh as much as the iron. Therefore the upthrust is not great enough to balance the weight of the iron. But when the iron is shaped into the form of a *ship* it displaces more water than it could before it was so shaped. In fact, the ship may displace its own weight and that of the very small weight of air inside it before it is even half submerged.

A special type of ship is the *floating dry dock*. This has several watertight compartments which can be filled with water or air as required. When the compartments are filled with water the weight of the dock is increased. The

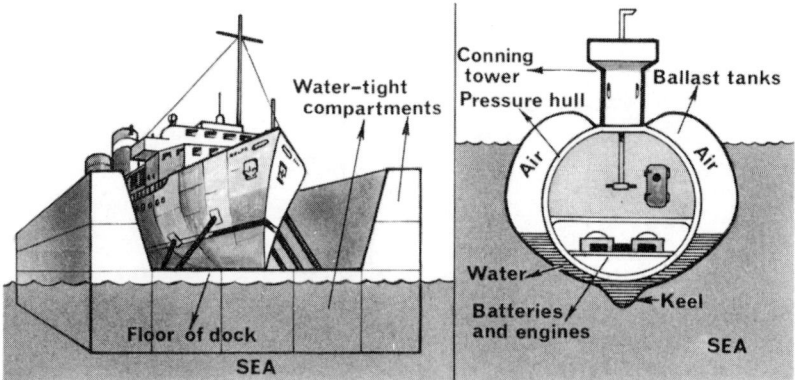

Figure 11.9. A floating dry dock. How is the ship returned to the water after the repairs are finished?

Figure 11.10. A cross-section through a submarine showing the position of the water ballast tanks that control its depth.

upthrust needed to balance this increased weight is then greater, and thus the dock sinks lower in order to displace a greater volume of water. The 'floor' of the dock sinks below the surface of the surrounding water. Any vessel needing repairs to its hull is floated into the dock and secured. Air is pumped into the compartments to force out the water and the dock rises until the 'floor' becomes dry so that the repairs can be carried out.

A *submarine* can float on the surface of water or can sink below the surface as desired. The same principles apply as in the case of the floating dock. When it is necessary to submerge, water is taken into the ballast tanks until the weight of both the submarine and the water in the ballast tanks together is greater than the upthrust on the submarine. When the submarine has to return to the surface the water is forced out of the ballast tanks by means of compressed air.

The depth to which a body floats is determined by the amount of liquid that must be displaced to make the upthrust equal to the total weight of the body. Some liquids are denser than others; for example, water is denser than paraffin, and mercury is much denser than water. Hence it follows that a body need not sink so deep in a denser liquid in order to obtain the necessary upthrust to float.

Density is the word we use to express the mass of a certain volume of a substance. The substance can be a solid, a liquid, or a gas.

$$\text{DENSITY (kg m}^{-3}\text{)} = \frac{\text{MASS (kg)}}{\text{VOLUME (m}^3\text{)}}$$

Density is commonly and conveniently expressed in $g\,cm^{-3}$ ($1\,g\,cm^{-3}$ $= 10^3\,kg\,m^{-3}$) because the mass of $1\,cm^3$ of water is $1\,g$ and therefore the density of water is $1\,g\,cm^{-3}$.

Often we wish to compare the density of another substance with that of water. We can do this by finding the mass (in grammes) of a certain volume (in cubic centimetres) of the other substance and hence determining its density. The comparison or ratio of its density to that of water ($1\,g\,cm^{-3}$) is then expressed as a number and is called the *relative density*.

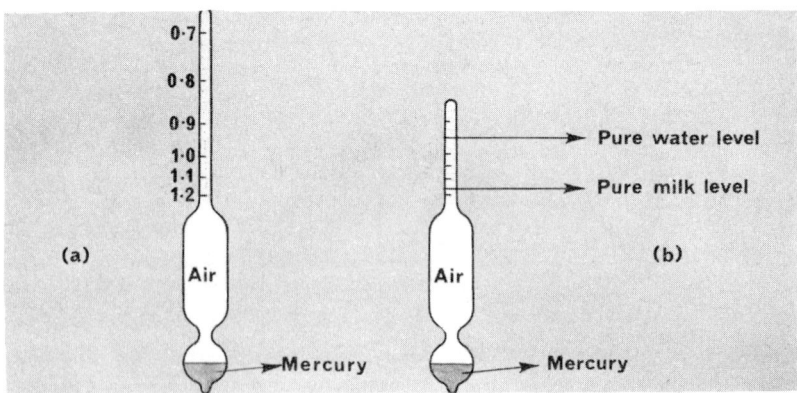

Figure 11.11. Hydrometers. (*a*) The scale of this hydrometer indicates directly the relative density of the liquid in which it floats. (*b*) The scale of this lactometer indicates the pure milk and pure water levels and a few divisions about these levels.

$$\text{RELATIVE DENSITY (ratio)} = \frac{\text{DENSITY OF SUBSTANCE (g cm}^{-3})}{\text{DENSITY OF WATER} (= 1 \text{ g cm}^{-3})}$$

Below is a table of the relative densities of some common substances.

Aluminium	2.7	Cork	0.22–0.26	Water	1.00
Iron	7.7–7.9	Wood (oak)	0.6–0.8	Methylated spirit	0.83
Brass	8.1–8.5	Ice	0.92	Paraffin	0.80
Lead	11.3	Rubber	0.15–1.7	Turpentine	0.87
Mercury	13.6	Glass	2.4–2.6	Glycerine	1.26

It is easier to float in sea water than in fresh water because the salt in the sea water makes it denser, and thus one can keep more of the head above the surface and so breathe more freely. A ship will also float higher out of sea water than it will out of fresh water. This sometimes presents problems in loading a ship so that it will not turn over in stormy weather. Often heavy ballast has to be taken on and put in the bottom of its holds to make it more stable.

The hydrometer is like a small ship of fixed mass designed to sink to various depths in liquids of different densities. It floats upright and has a stem on which a scale is marked. When it floats in water the surface of the water reads 1·0 on the scale of the hydrometer. A special hydrometer is used for testing the density of the acid in an accumulator. Another, known as a lactometer, by indicating the density gives a simple test of the purity and quality of milk. The lactometer can tell us immediately if the milk has been diluted with water.

Liquids in motion. When a liquid is driven along a straight horizontal pipe there is a progressive drop in pressure due to fluid friction and this can be measured in little side tubes. At the same time any constriction in the flow of the liquid causes a great increase in the speed of flow of the liquid and this, in turn,

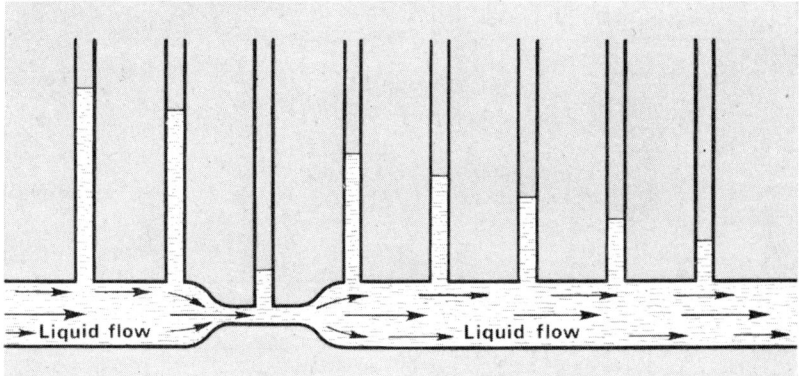

Figure 11.12. The changes in pressure of a liquid as it flows along a pipe that is constricted at some position. The pressure is observed in the side tubes that open freely into the main pipe and whose upper ends are open to the atmosphere.

produces a great drop in pressure in the side tube at the constriction. The pressure rises again when the tube widens and the flow speed is returned to normal.

Surface tension. The forces acting between molecules in a liquid cause those at or near the surface to act as if they formed part of a 'skin'. This property is called *surface tension*. On this surface 'skin' it is possible to float light objects such as iron filings, a needle, a razor blade, and a large piece of wire gauze. Some insects, like the pond skater, 'walk' about on the surface of the water and are extremely agile in doing so. Mosquito larvæ hang their breathing tubes on this 'skin'.

Forces between like molecules, that is, molecules of the same substance, account for the property known as *cohesion*, and forces between unlike molecules for the property of *adhesion*.

Surface tension causes light soap bubbles and small water drops to be round. To make lead shot, molten lead is passed through a fine sieve held high in the air. As the lead falls freely it breaks up into drops because of the surface tension, cools, and solidifies into perfect spherical shapes.

The force of attraction between the molecules of water and glass causes water to run up the sides of a piece of glass immersed in water. It wets the glass. This property is known as *capillary attraction*.

During dry weather the small quantity of moisture in the soil can be conserved for use by the plants if the farmer thoroughly breaks the top one or two inches of surface soil. The reason is that a hard compact surface soil has fine spaces or tubes between the grains of soil and these allow capillary attraction to take place so that the moisture rises to the surface and there evaporates. Breaking up the surface soil destroys the fine capillary tubes and only leaves wide ones up which moisture cannot travel.

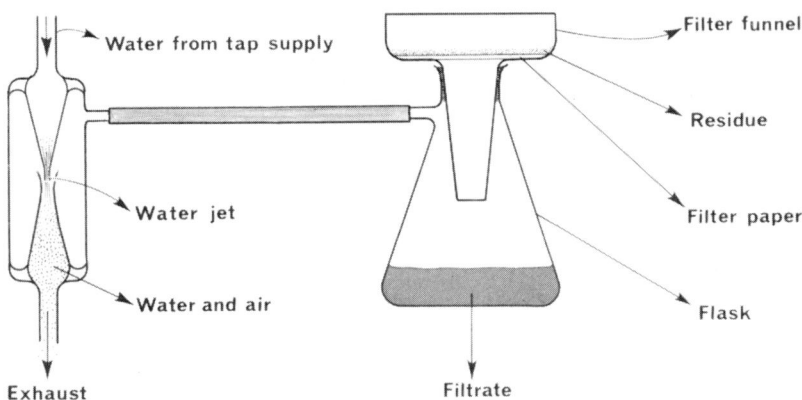

Figure 11.13. A water jet filter pump. The drop in pressure around the jet causes air to move into the water jet and thus exhaust the flask. Why does the filtrate pass through the filter paper?

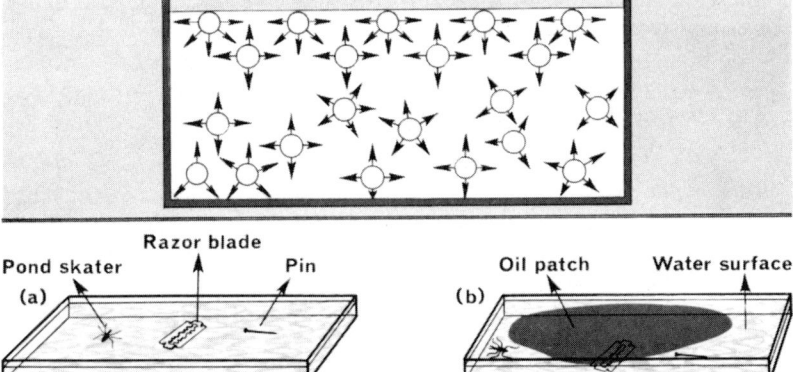

Figure 11.14. The forces on the molecules near the surface of a liquid are greater downwards and sideways than they are upwards, and thus the molecules tend to cling together and prevent the surface layer being broken. Forces acting on those molecules inside the liquid are balanced in every direction.

Figure 11.15. (*a*) Forces due to cohesion between water molecules form a 'skin' at the surface layer on which light objects, like pins and mosquito larvae, can float, and insects, like the pond skater, move about. (*b*) A few drops of oil poured on the water surface spread out in a thin film because the adhesion forces between the oil and water molecules are greater than the cohesion forces between the oil molecules themselves. The small objects sink, the mosquito larvæ drown, and the insects move away quickly from the advancing oil film.

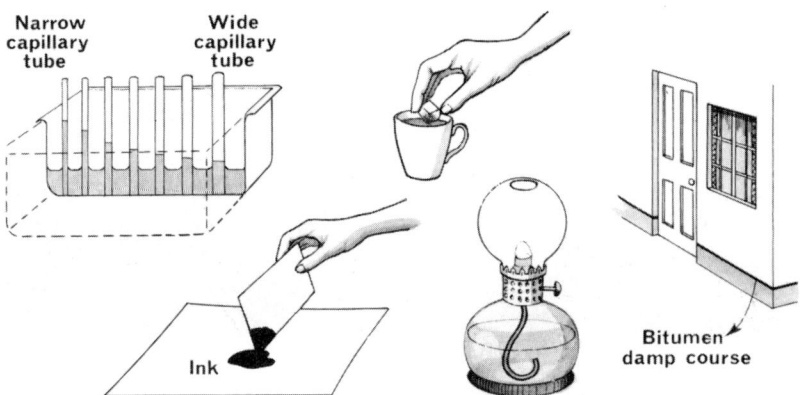

Figure 11.16. Capillary attraction causes water to rise up the bore of these glass tubes. Notice that it rises further up the tubes of narrow bore. It also causes ink to run into the piece of blotting paper, tea to soak into the lump of sugar, oil to rise in the wick of an oil lamp, and moisture to creep through the walls of a house as far as the 'damp course' of slate, 'blue bricks', or bitumen-coated felt.

Concrete floors laid directly on the earth sometimes feel cold and damp because moisture rises up to the surface from the earth below through the fine capillary tubes that form the pores of the concrete. This should be prevented by laying the concrete on a few inches of cinders and clinkers that have pores too large to allow capillary attraction to take place.

Pressure, depth, and density. The pressure in a liquid varies because of both the depth and the density as can be shown by the following reasoning. Consider a column of liquid vertically above a certain area. The pressure on that area is according to the definition on page 73 represented by:

$$\text{PRESSURE} = \frac{\text{THRUST due to the liquid above}}{\text{AREA on which the liquid stands}}$$

$$= \frac{\text{FORCE of attraction of gravity}}{\text{on the liquid directly above the area}}$$
$$\overline{\text{AREA}}$$

$$= \frac{\text{VOLUME} \times \text{DENSITY} \times g}{\text{AREA}}$$

$$\therefore \underline{\text{PRESSURE} = \text{DEPTH} \times \text{DENSITY} \times g}$$

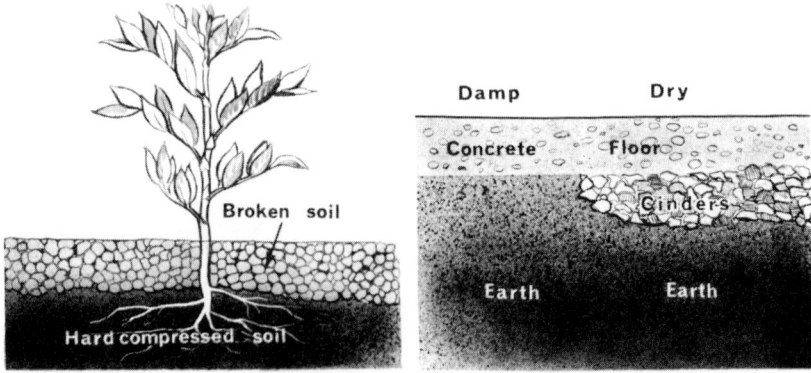

Figure 11.17. Water from some considerable depth below the surface of the soil rises by capillary attraction through the fine pores between the soil particles and reaches the root hairs of plants where it is absorbed. 'Dry farming' helps to prevent the evaporation of moisture by breaking up the fine pores near the surface of the soil.

Figure 11.18. One section of this concrete floor is cold and damp and is said to 'sweat'. The other is dry because it has the proper foundation of cinders and clinkers.

Questions on chapter 11

1. Define pressure. Describe an experiment to show that the pressure in a liquid increases with depth.

Oil and brine are placed in opposite arms of a U-tube. The level of the oil is 8.4 cm, and that of the brine is 6.8 cm, above the oil-brine dividing level. If the density of the brine is 1.05 g cm^{-3}, calculate the density of the oil.

2. Draw a diagram to illustrate an experiment showing that pressure in a liquid acts in all directions.

3. Why is it impossible for divers to descend to great depths?

4. Why is a hole in a ship near the bottom more dangerous than a similar hole nearer the surface of the water?

5. Why does a more powerful jet of water come from the cold tap than from the hot tap in most houses?

6. What part does a water tower play in the distribution of water to houses in a town?

7. Does the thrust on a lock gate depend on the horizontal area of the water surface in the canal above the gate? Explain your answer.

8. What are the advantages of hydraulic machinery?

9. Explain how hydraulic brakes on a motor-car are equalized. Are the braking forces on the front and rear wheels the same?

10. Car service stations are sometimes equipped with hydraulic hoists. Draw a diagram to show how these work.

11. A cube of material has a side of length 10 cm and a mass of 800 g. What is the density of the material?

If it were placed in water, what mass of water would be displaced? What volume of the cube would be submerged?

12. A cube of material of density 0.75 g cm^{-3} has sides 10 cm long. What is its mass?

What mass of liquid will it displace when floating in a liquid? What volume of the liquid will be displaced if the density of the liquid is 1.25 g cm^{-3}?

13. Define (a) density, and (b) relative density. How are these two quantities related?

A block of wood, of density 0.65 g cm^{-3} has a mass of 13 g. What is its volume? When a piece of metal of mass 9 g is attached to the block of wood, their combined density is the same as that of water. Find the relative density of the metal.

14. A bottle has a mass 25.6 g when empty and 49.1 g when full of water. What volume of alcohol would it hold? Calculate the mass of the bottle when it is full of a liquid of density 1.2 g cm^{-3}.

15. A large stone has a mass of 240 kg. When it is hanging freely in fresh water its apparent mass is only 80 kg. Why is there this apparent loss? What is the volume and the density of this stone?

16. An empty bottle has a mass of 20 g. When filled with water the total mass is 120 g; when filled with brine the mass is 135 g. What is (a) the volume of the liquid held in the bottle, and (b) the density of brine?

17. If the density of ice is 0.9 g cm^{-3}, calculate the proportion by volume of the ice which is below the surface when it is placed in pure water.

18. State the principle of Archimedes. Describe briefly how you would verify this statement experimentally.

A piece of rock 'weighs' 15.4 g in air and 9.8 g in a liquid of relative density 0.8. Calculate the relative density of the rock.

19. Describe a simple hydrometer and explain how to use it to find the relative density of a suitable liquid.

20. Why is it difficult to write clearly with pen and ink on newspaper or on oiled paper? How is this difficulty overcome in paper prepared for writing with ink?

21. Why is sawdust or sand sometimes used to clear oil from the floor of a garage?

22. How are concrete floors laid so that they remain dry even in wet weather?

23. Lead shot is made by pouring liquid lead through fine holes placed at a certain height above a tank of cold water. Why is the shot spherical when it is collected from the water?

24. Why does wet hair cling together and stick to your head while dry hair waves about in the breeze?

25. All other factors being equal, would a toothpaste with a low surface tension be better than one with a high surface tension?

26. Why do some motor-car radiators that normally hold water perfectly well begin to leak when an anti-freeze is added in the winter?

12. Some properties of gases

A gas behaves in many ways like a liquid. Its pressure varies with depth and is transmitted in the same way as it is in a confined liquid. A gas possesses the properties necessary to cause buoyancy and flotation. A gas in motion, again like a liquid, creates a low pressure region around a tapered jet. The one important difference between gases and liquids is that a gas is compressible whereas a liquid cannot be compressed. A gas occupies the space of its container, however large or small this may be, whereas a liquid has a fixed volume.

Atmospheric pressure. Air has weight and the atmosphere extends upwards from the surface of the earth, getting rarer and rarer for at least 320 km. The air presses down on the surface of the earth with a pressure equal to about 1×10^5 N m^{-2} (called a *bar*), or with about the same pressure as a column of water 10 m high or a column of mercury 76 cm high.

We hardly notice the pressure of the atmosphere on our body because the pressure acts in all directions, for example it presses on our cheeks from the outside and also from the inside.

There are many cases where the pressure of the air is found to be useful.

The pressure of the atmosphere causes the liquid to be pushed up into the devices shown in the first four illustrations of Figure 12.2 when the pressure of the air above is reduced by one of the ways shown. Why does the sucker stick to the smooth windscreen of the car? Why should two holes be punched in the

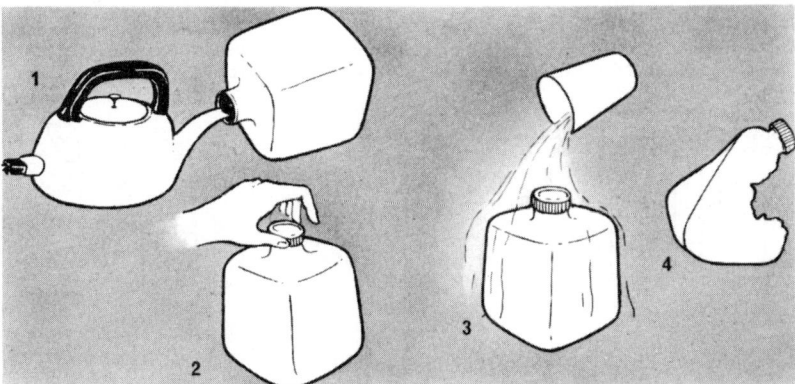

Figure 12.1. An experiment to show the effect of the pressure of the atmosphere on the sides of an empty vessel. (1) Steam from a boiling kettle of water is directed by the spout into a polythene water container. (2) After most of the air has been forced out of the container and replaced by the steam the container is removed and its cap screwed on firmly. (3) Some cold water is sprinkled on the outside of the container and the steam inside condenses forming a partial vacuum. (4) The atmospheric air pressure crumbles the container as it would a paper bag.

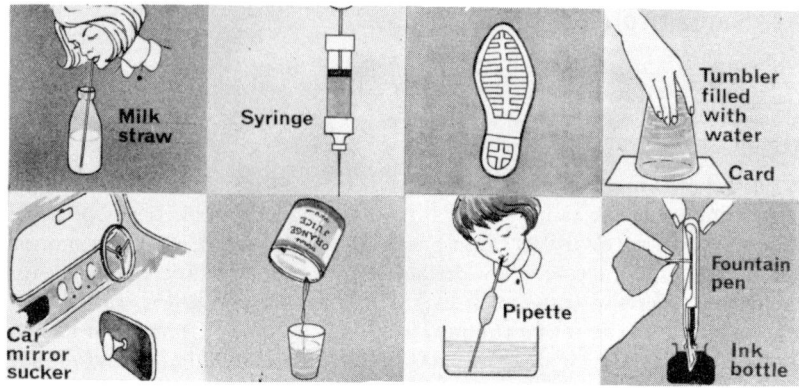

Figure 12.2. Examples of devices that rely on atmospheric pressure.

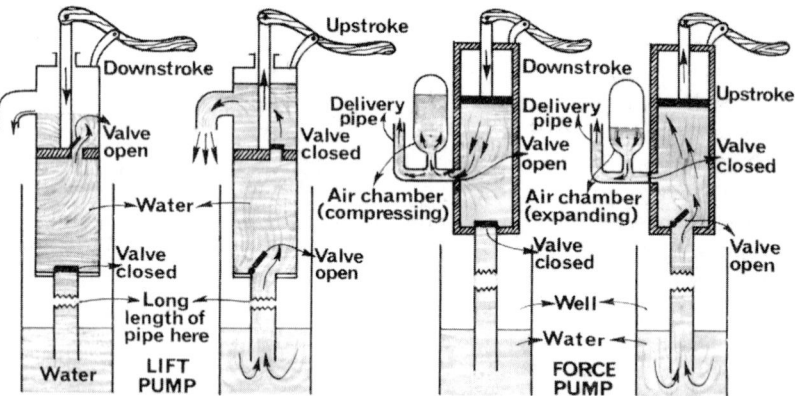

Figure 12.3. The lift pump and the force pump in action. The arrows show the movements of the water during each stroke of the pistons. Within what vertical distance must both pumps be mounted from the surface level of the water in the well?

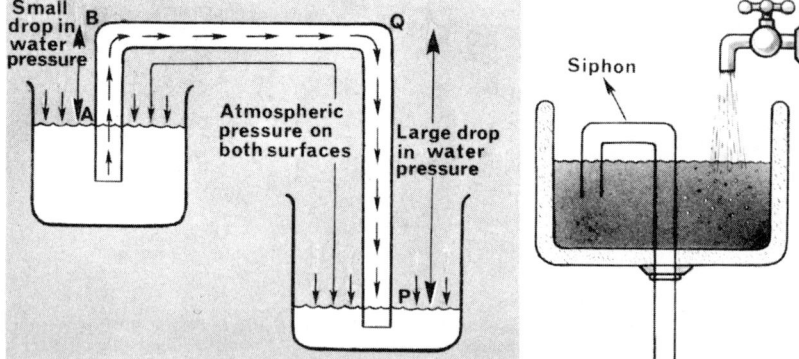

Figure 12.4. A siphon transferring liquid from the higher vessel to the lower one. How is the liquid made to start flowing and how can it be stopped quickly?

Figure 12.5. An automatic intermittent flushing tank. When does this start to flush?

can of orange juice? Why does this rubber-soled sports shoe not slip on a wet smooth surface? Why does this card remain pressed against the tumbler even though it is turned upside down?

Pumps. The atmospheric pressure plays a vital part in the working of a lift pump used for raising water from a well. During the upstroke of the piston a reduction of pressure takes place under the piston and the atmospheric pressure forces water from the well up the tube into the cylinder. The maximum theoretical height to which this can take place is 10 m, but in practice, due to leakages in the valves and between the piston and the cylinder wall, the limit is probably nearer to 7 m. In the case of a deep well this height is insufficient to bring water to the surface so a force pump has to be used.

The water emerges from the spout of the lift pump unevenly but harmlessly as the height of the water above the piston varies with the position of the stroke, but in the case of the force pump the water is delivered in definite jerks that can easily shake apart pipe connections. It is much better to keep a steady flow of water rising up the delivery pipe of the force pump because starting and stopping such a great height of water would cause a great strain on the pump. So an air chamber is included in the delivery pipe to act as a cushion.

A siphon consists of a bent tube, open at both ends, one arm of which is longer than the other. It enables us to empty the liquid from a vessel at a higher level to another at a lower level without the use of a tap and without pouring.

The siphon operates because the pressures are unbalanced. The atmospheric pressures at A and P are the same. The pressures at B and at Q are less than atmospheric and different from one another because of the weight of the liquid in the two unequal arms AB and PQ of the siphon. Because AB is shorter than PQ the pressure at B is greater than the pressure at Q, and hence the liquid at B moves towards the lower pressure liquid at Q and continues until the higher vessel is emptied. The rate of flow increases as the difference between the lengths of the two arms increases.

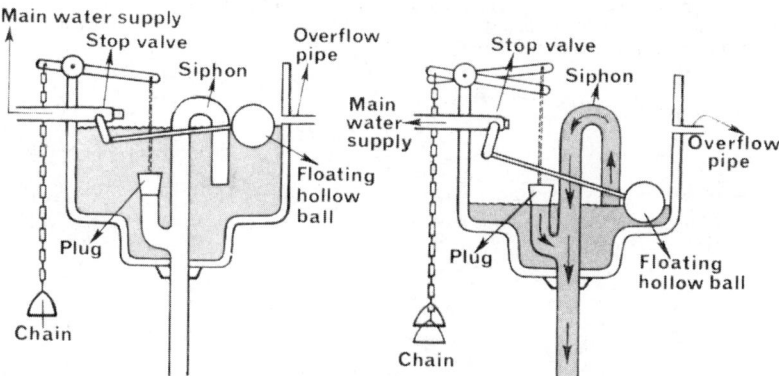

Figure 12.6. A lavatory flush tank. How is the flow started and why does it continue?

G

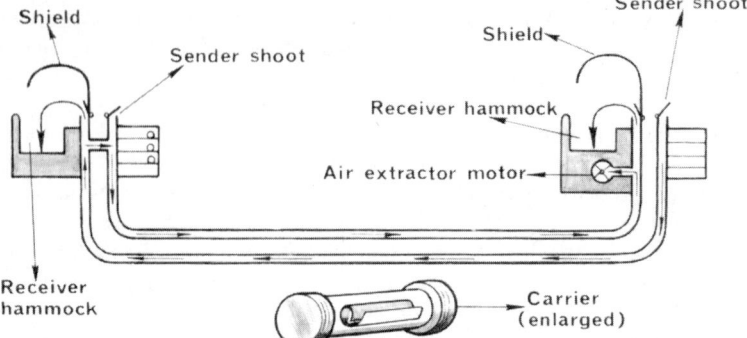

Figure 12.7. An air-driven (pneumatic) carrier system for small messages and money. The flap over the sender shoot is opened and the carrier, with its messages sealed inside, is inserted. The pressure of the air pushes the carrier along the pipe, until at the receiver end the carrier, then moving at speed, hits and opens the light flap where it is guided quietly by the curved shield into a padded hammock.

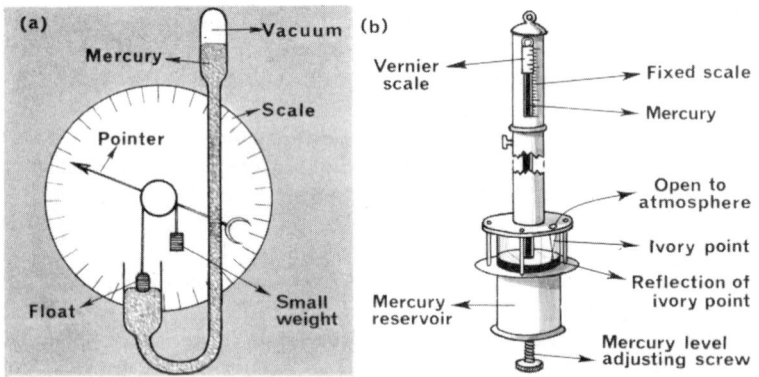

Figure 12.8. Two types of mercury barometer. (*a*) The atmospheric air presses on the mercury in the open end of the U-tube and this balances the pressure of the mercury in the closed end. (*b*) The refined model, called a Fortin barometer, enables very accurate readings to be observed—the ivory pointer and its reflection in the mercury are used to adjust the mercury to a fixed level in the lower open end, and a vernier is used for reading the height of the mercury near the upper closed end.

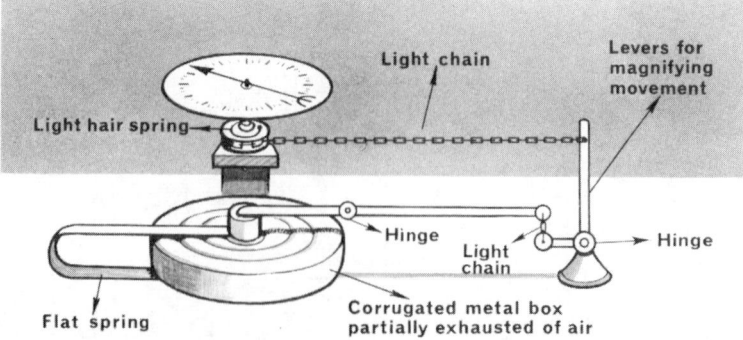

Figure 12.9. An aneroid barometer. This can be made small and portable. Can you see how the levers and the wheel and chain magnify the very small movements of the surface of the metal box to give a big movement of the end of the pointer over the scale?

A barometer measures the pressure of the atmosphere usually on a scale representing the height in centimetres of a column of mercury producing the same pressure. One type of barometer actually balances the pressure of a column of mercury against the air pressure, and another measures how far the air pressure is able to crush a corrugated metal box that is nearly exhausted of air and then indicates this measurement in terms of the height of a column of mercury.

An important use of the barometer is for measuring the height of an aircraft above the ground. We know that for places not far above sea level the decrease in atmospheric pressure is about 1 centimetre of mercury for every 108 metres rise in height. For the convenience of pilots a very sensitive aneroid barometer is made and this is called an *altimeter*. Its readings are stated directly in metres. Of course the pilot always has to keep the altimeter in correct adjustment as the true atmospheric pressure at sea level changes from place to place and from hour to hour as he flies. Provided the pilot knows exactly the correct value of the atmospheric pressure at sea level where he is at the time he is there, he can adjust his altimeter to give him his correct altitude.

Buoyancy in gases. Every body in air has an upthrust acting on it due to the weight of the air it displaces. A balloon filled with a gas lighter than air, either hydrogen or helium, displaces a weight of air greater than the sum of its own weight and the weight of the gas inside it. The difference between the weight of air displaced and the sum of these two weights gives the carrying capacity of the balloon. In this condition the unladen balloon must be held down by ropes to prevent it rising. Once freed the balloon will rise until it reaches less dense air

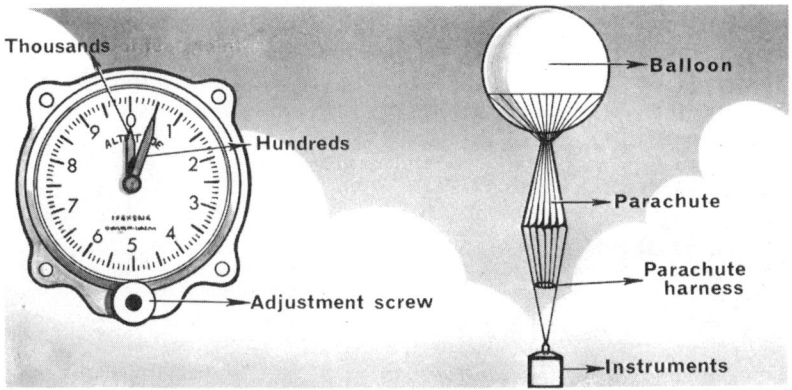

Figure 12.10. An altimeter as fitted into the cockpit of an aircraft. The pointer and the figures are painted so that they glow in the dark and thus enable the pilot to see them at night.

Figure 12.11. A weather balloon rises carrying with it instruments for recording temperature, humidity, and air pressure. Eventually at a predetermined height the balloon bursts and the instruments gently fall to the earth suspended on the parachute folded immediately below the balloon. What slows the rate of fall of the parachute?

high above the earth's surface, where the upthrust is less and just equals the weight of the balloon and its gondola. The balloon then 'floats' steadily in its atmosphere of rarer air.

Gases in motion. The low pressure produced by rapidly moving streams of gas have many useful applications. Sprays for paints, scents, and disinfectants use air jets and the Bunsen burner uses a jet of inflammable gas.

The wing of an aircraft (*aerofoil*) is so shaped that the air is driven past its upper surface very rapidly reducing the pressure there, whereas against the lower surface the air slows down and builds up an increased pressure. These two different pressures result in the 'lift' given to the wing making flight possible for 'heavier than air' machines.

Compression and expansion of a gas. If we close the outlet of a bicycle pump with a finger and then force the piston inwards we shall observe that it becomes more and more difficult to compress it as the piston goes further inwards. In fact, if you wish to halve the volume, the total pressure needed on the piston, made up of the atmospheric pressure and the pressure due to your effort, will have to be doubled. If the volume is to be reduced to one third then the total pressure needed must be trebled. There is, however, a precaution that must be considered in this simple relationship and that is that the temperature must remain constant during the changes. Robert Boyle stated these observations, that are true for all gases, in a law that bears his name. It states that the pressure of any gas trapped in a container multiplied by its volume remains constant if the temperature is unchanged, or expressed in the form of an equation it is:

$$\text{PRESSURE} \times \text{VOLUME} = \text{CONSTANT}$$

There are many useful applications of gases under pressure that are found in daily life. All of them depend on the tendency of a compressed gas to expand as the pressure is reduced and contract again as the pressure is once more increased. The compressed air inside a motor-car tyre causes it to move in and out, smoothing out the irregularities of rough roads, as stones and pot-holes vary the outside pressure on the tyres.

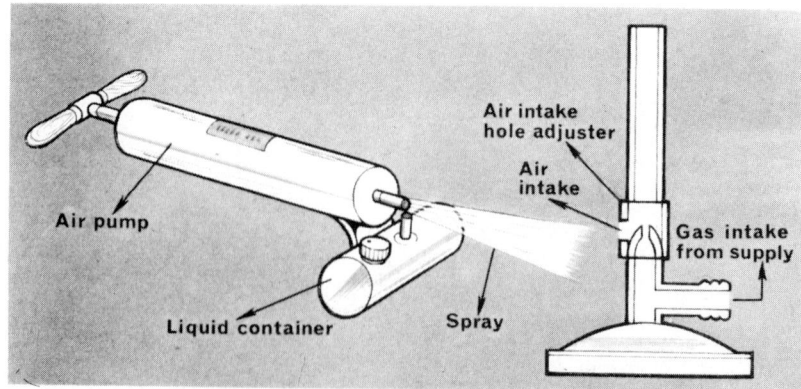

Figure 12.12. Two uses of a gas jet—a disinfectant spray and a Bunsen burner.

The air pressure inside a motor-car tyre varies considerably according to the make of tyre and whether the car is designed for high-speed touring or for carrying big and heavy loads. The tyre on an average passenger car is inflated to about the pressure of 3 atmospheres which means about three times the normal atmospheric pressure. Hence the pressure in the tyre is 2 atmospheres above the air pressure outside, and this pressure of 2 atmospheres is read on the tyre air pressure gauge.

The normal atmospheric pressure is called a *bar* (see page 85), which is $1.0 \times 10^5 \text{ N m}^{-2}$, and meteorologists speak in terms of millibars (10^{-3} of a bar) when referring to barometric changes.

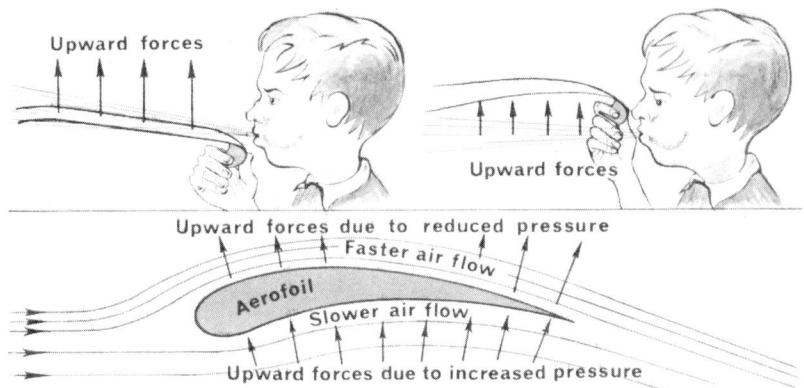

Figure 12.13. The boy shows how a light curved strip of paper can be raised by blowing either above it or below it. Similar forces on a well-shaped aerofoil produce the 'lift' needed to keep an aircraft in flight.

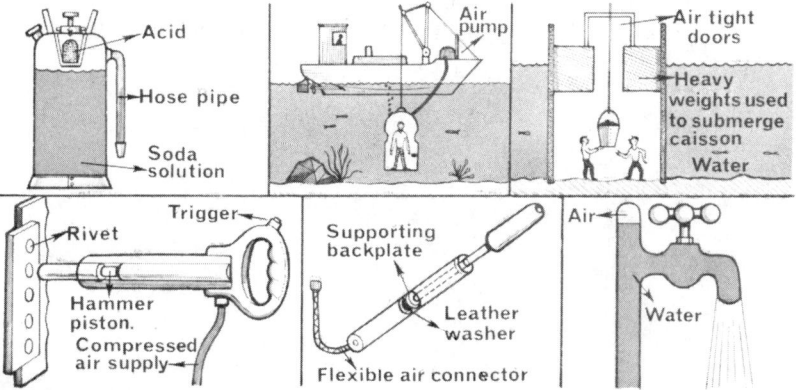

Figure 12.14. Some devices that make use of compressed gases—the fire extinguisher, the diving bell, the pneumatic caisson, the compressed air riveter, the bicycle pump, and the air chamber that stops 'hammering' when a cold water tap is turned off suddenly.

Questions on chapter 12

1. Which has the greater weight—an inflated football or the same football deflated? How would you prove your answer?

2. Why is it better to punch two holes, on opposite edges of the top of a tin can, in order to pour out the liquid inside, than one single hole even if that one is larger than the others?

3. Why does ink flow into the rubber sack of a fountain pen when it is being filled?

4. Describe how you would make a simple barometer. Include in your account the details of the tube you would use, how to fill it, and how to use the barometer to measure the pressure of the atmosphere.

What will be the effect on the height of the mercury column if a small amount of air is introduced into the space above the mercury?

What would be the effect on the height reading if you took the barometer (a) to the top of a mountain, and (b) to the bottom of a mine?

How does an aneroid barometer differ from a mercury barometer?

5. How does a vacuum cleaner draw up the dust from the carpet? Why are large nails often left undisturbed by the cleaner?

6. Why can a force pump drive water higher than a suction or lift pump can lift it?

7. With what gas or vapour is the space above the mercury in a barometer filled and why?

8. If a barometer tube, reading correctly, with mercury in it, is tilted, what happens in the space above the mercury?

9. Is there any limit to the height over which water can be siphoned?

10. An air compressor in a garage has a reserve cylinder of volume 3 m³ and the air in it has a pressure of 4.2×10^6 N m^{-2}. If this air is used for inflating motor-car tyres, what volume can be supplied at a pressure of 3.0×10^5 N m^{-2} before the pump has to start working to refill the cylinder? Assume that the pump starts when the pressure falls below 3.0×10^5 N m^{-2} in the reserve cylinder.

11. A motor-car tyre is inflated to a pressure of 1.5×10^5 N m^{-2}. It supports a load of 4.5×10^3 N. What area of the tyre is in direct contact with the ground?

13. Work, energy, power

The scientific meaning of the word 'work' differs from that in ordinary use. For instance, a scientist does not say that a person is working who is trying to solve a problem in mathematics, or who holds up a fence by pressing against it as it is being fixed in position. In the scientific sense he is not doing any work at all. He does work when he lifts a heavy package a distance against the force of attraction of gravity on the package. See Figure 13.1. Work is done only when a force moves an object through a distance. The force can be a push or a pull; it can be applied directly by a person or an engine or, as we shall see later, indirectly with the aid of a machine. No amount of force will do work unless the force moves its point of application. In fact, the work done is the product of the force and the distance moved in the direction along which the force acts, or expressed in the form of an equation is:

$$\text{WORK} = \text{FORCE} \times \text{DISTANCE}$$

Work is measured in units that are the product of the units of a force and a distance. The unit of force is a newton and the unit of distance is a metre. Therefore the unit of work is a newton metre and this is often expressed by the symbols N m. As this is such an important unit the newton metre is called a *joule* (J) in honour of the physicist who made a great contribution to our knowledge of work and energy. Work and energy are both measured in the same unit, a joule, as we shall learn below.

Of the men shown working in Figure 13.1 two are storing up the work they do—one in placing the heavy case in a high position above the ground and the

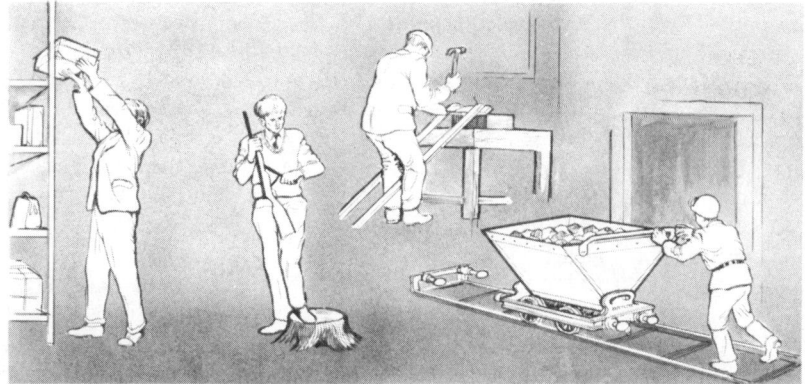

Figure 13.1. All these men are doing work. One is lifting a heavy case against the force of gravity, another is compressing the spring in his air gun, another is pushing a loaded truck of earth along a railway, and the fourth is hammering a nail into a piece of wood.

other in compressing the spring in his air gun. They can release this 'stored work' when they wish. The bodies in which this work is stored have now got a new property called *energy*, that is they have now acquired a new *capacity for doing work*. The energy possessed by the heavy case on the top rack can do work if it is allowed to fall; for instance, it can squash the toe of the man if it hits his foot. The energy of the loaded spring can shoot a pellet a great distance when the trigger is pressed and the pellet could pierce a hole in a piece of wood placed in its track. These are examples of *potential energy*—one kind of mechanical energy—due to the position or condition of the body.

The energy possessed by the moving truck of earth is *kinetic energy*—another kind of mechanical energy—due solely to the motion of the body. When the truck hits the buffers at the end of the track it will lose its kinetic energy, compress the springs in the buffers, and create potential energy. If, however, the brakes are applied to stop the truck the brakes will get hot because the kinetic energy of the moving truck will turn into heat energy. The man hammering the nail into the wood is converting some chemical energy stored in the food in his body into the kinetic energy of the moving hammer head. This kinetic energy is then converted into heat energy created by the friction of the nail as it is pushed into the wood and the nail becomes hot.

Potential energy and kinetic energy are both forms of mechanical energy.

In our further studies of science we shall encounter other forms of energy such as thermal energy, electrical energy, and chemical energy. We shall also learn that energy cannot be created or destroyed—it can only change from one form to another. As energy is the capacity a body possesses for doing work, any change in that energy must be equal to the work it does, and is therefore measured in the same units of joules.

The *rate at which work is done* is termed *power*; power is the amount of energy changed every unit of time. Power is therefore represented by the equation:

$$\text{POWER} = \frac{\text{WORK DONE}}{\text{TIME TAKEN}}$$

and is measured in units of the unit of work per unit of time, that is joule per second ($J s^{-1}$) which is called a watt (W) after the well-known Scottish engineer. This is a small unit when we consider the power of an engine and thus a larger unit of 746 watts is used and this is called a *horse-power* (h.p.).

It is necessary to know the horse-power that various types of engines and electric motors can develop in order that they may be selected for the task they have to do. The horse-power is usually measured by finding the work it can do in a certain time against a brake and it is thus sometimes called the brake horse-power (b.h.p.). It may be many years before the power of an engine is reckoned in watts instead of horse-power.

The engineer working in Figure 13.2 shows how this can be done in the case of a small electric motor but similar operations can be carried out with larger engines and even turbines. The motor is made to do work against the friction developed between a belt and a rotating drum. The belt drags down the left-hand spring balance (f_1) and releases the right-hand spring balance (f_2). The difference in the readings of the two spring balances gives the braking force. The distance travelled by this force is determined by the circumference of the drum times the number of revolutions read on the counter in the engineer's left hand. The time is read on the stop watch in his right hand.

The power equation above can be rewritten in this way:

$$\text{POWER} = \frac{\text{FORCE} \times \text{DISTANCE}}{\text{TIME}},$$

$$\text{POWER (in N m s}^{-1}) = \frac{\substack{\text{Difference in forces} \\ \text{on brake bands} \\ \text{(in N)}} \times \substack{\text{Circumference of} \\ \text{brake pulley wheel} \\ \text{(in m)}} \times \substack{\text{Number of} \\ \text{revolutions} \\ \text{observed}}}{\text{Time of observation (in s)}}$$

During one test by the brake method to find the brake horse-power (b.h.p.) of a small electric motor these readings were observed:

$$\text{Reading of spring } f_1 = 50 \text{ N}$$
$$\text{Reading of spring } f_2 = 20 \text{ N,}$$
$$\text{Circumference of brake drum} = 0.3 \text{ m}$$
$$\text{Number of revolutions} = 2\,480,$$
$$\text{Time of observation} = 60 \text{ s.}$$

The power is calculated by converting these readings to the proper units and putting them in the equation:

$$\text{Power} = \frac{\text{Force (N)} \times \text{Circumference (m)} \times \text{No. of revolutions}}{\text{Time (s)}},$$

$$\therefore \text{Power} = \frac{(50 - 20) \times 0.3 \times 2\,480}{60} \text{ N m s}^{-1}(\text{W})$$

$$\therefore \text{b.h.p.} \atop \text{of the motor} = \frac{30 \times 0.3 \times 2\,480}{60 \times 746}, \text{ (1 h.p.} = 746 \text{ W)}$$

$$= 0.5.$$

$\therefore$ the small electric motor under test has a power of $\frac{1}{2}$ b.h.p.

Figure 13.2. The brake method of finding the brake horse-power (b.h.p.) of a small electric motor. The stopwatch and the revolution counter enable the engineer to determine the number of revolutions per second of the motor when it is working continuously against the braking force as measured by the two spring balances f_1 and f_2.

Questions on chapter 13

(1 h.p. = 746 W, take g to be 10 m s^{-2}).

1. What is meant by force, work, and power?

A lift weighing 200 N is used to raise a load of 800 N through a height of 20 m in 10 s. What is: (a) the useful work done; (b) the total work done; (c) the efficiency of the lift; and (d) the horse-power required to raise the lift and its load at this rate?

2. In which of the following situations is work being done: (a) a hungry boy eating an apple; (b) a girl playing a piano; (c) a boy keeping a door shut against a strong wind; (d) a cat climbing a tree; (e) a porter standing still with a case on his shoulders; (f) an elephant pushing over a tree; and (g) milk pressing against the sides of a churn?

3. At a sports meeting two boys of exactly the same height both clear 1.45 m in the high jump. Explain why one boy did 20% more work in doing this than the other boy did.

4. A ball starts from rest and rolls downhill; at the bottom of the hill it hits a wall. Use this illustration to explain the terms potential energy and kinetic energy.

5. Two boys ride a motor-scooter to the top of a hill which is a vertical climb of 66 m in 180 s. The boys and the scooter weigh altogether 2 500 N. What work does the engine of the scooter do against the force of gravity? What horse-power does the engine deliver during this climb?

6. Explain why the motor-car with the greatest acceleration is not necessarily the most powerful one.

7. What is meant by (a) energy, and (b) power?

Water is pumped to a height of 40 m at the rate of 0.01 m^3 s^{-1} by a motor-driven pump. If the efficiency of the system is 80%, what horse-power must be supplied by the motor?

8. A boy weighing 500 N climbs a hill 33 m high in 10 min. What work does he do during the climb, and what horse-power does he develop?

9. The reservoir feeding a power station with water is 1 100 m above the turbines. If the turbines convert all the energy of the water supplied to them into useful work, what horse-power should be generated by the turbines when they draw 6 000 kg of water every second?

10. An express train is moving at 100 km h^{-1} and the engine is developing 6 000 b.h.p. What force is it exerting on the train? What forces are acting on the wheels and the track? Express your answers in newtons.

11. Explain why two horses can often pull out a large powerful motor-car when it gets stuck in mud or sand, whereas the car engine cannot be made to move the motor-car.

14. Machines

A machine is a device that enables us to do work or to apply a force more easily and conveniently. This does not mean that a machine does more work than is put into it. On the contrary the useful mechanical work done by a machine is always less than the mechanical work put into it because some of the work is transformed into heat energy because of the friction of the moving parts.

The lever, pulley, wheel and axle, inclined plane, screw, and wedge are all simple machines.

There are three reasons why machines are used in spite of the loss of useful mechanical energy:

1. To obtain a large force by applying a small force, e.g. a motor-car jack.

2. To move a smaller force through a greater distance by moving a larger force through a shorter distance, e.g. a bicycle.

3. To change the direction in which the applied force acts, e.g. a single fixed pulley.

The useful work done by a machine is called the *output* and the work done on the machine is called the *input*. If there were no friction then the output would equal the input and the machine once started would be in a state of perpetual motion. Such perpetual motion machines do not exist because some work must always be lost as heat owing to the friction of the moving parts.

The resistance that has to be overcome when a machine does work is called its *load*, and the force applied to the machine to enable it to do this work is called the *effort*.

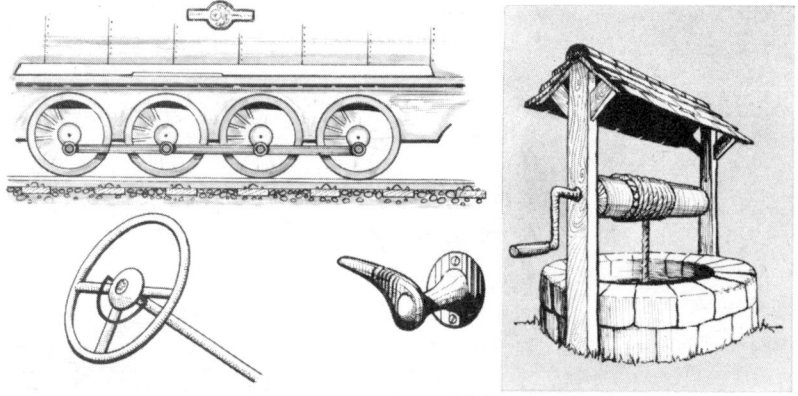

Figure 14.1. Various types of wheels and axles in common use. In what respect is the set of driving wheels on a locomotive different from the steering wheel, the door handle, and the windlass?

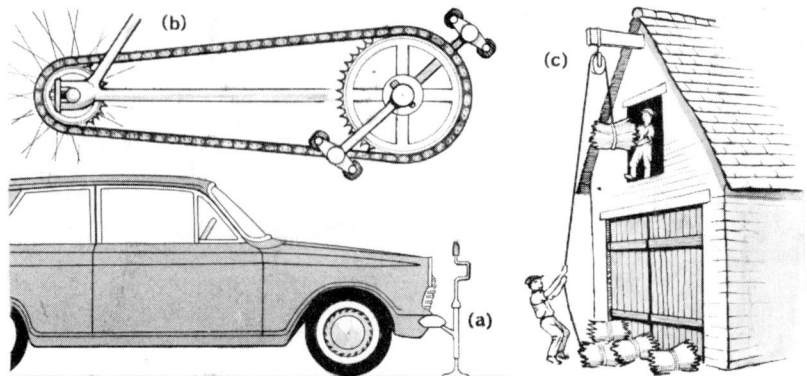

Figure 14.2. (*a*) The car jack enables one person to raise a heavy car. (*b*) The gear wheels and chain of a bicycle enable a person to move great distances whilst his feet cover much smaller distances. (*c*) The single fixed pulley makes it more convenient but not easier for a person to raise a body because he still has to exert the same force or even a slightly greater one for some force is needed to bend rope.

Figure 14.3. The inclined plane being used as a machine to raise a heavy weight off the ground. In both cases the effort to run the barrel or car up the ramp is much less than the vertical force that would be needed to lift it.

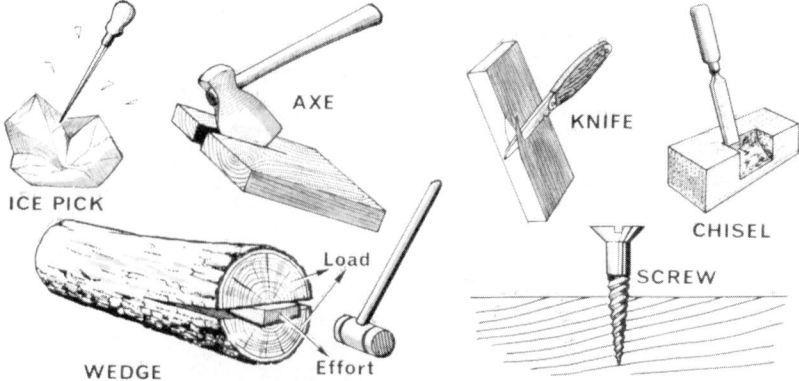

Figure 14.4. Six examples of a wedge being used as a machine. The directions of the load and effort are shown on one of the drawings. The wedge is really a double inclined plane and the thinner the wedge the tougher the log it can open.

The *mechanical advantage* (M.A.) of a machine is a number that expresses how much greater the load is than the effort and is given by:

$$\text{MECHANICAL ADVANTAGE} = \frac{\text{LOAD OVERCOME}}{\text{EFFORT APPLIED}}$$

The mechanical advantage may be greater than 1 as in the case of a crow-bar; equal to 1 as in the single pulley; or less than 1 as in the bicycle. Sometimes the convenience of the machine is more important than the gain or loss in force.

The ratio of the distance moved by the effort to the distance moved by the load in equal times is known as the *velocity ratio* (V.R.) of the machine.

$$\text{VELOCITY RATIO} = \frac{\text{DISTANCE MOVED BY EFFORT}}{\text{DISTANCE MOVED BY LOAD}}$$

By using the relationship, Work = Force × Distance, it is possible to calculate the output of the machine.

Output (Work) = Load overcome (Force) × Distance moved by load.

or,   OUTPUT = LOAD × DISTANCE MOVED BY LOAD,

and by a similar argument:

INPUT = EFFORT × DISTANCE MOVED BY EFFORT.

The *efficiency* of the machine is the ratio of these two values—the output and the input:

$$\therefore \text{EFFICIENCY} = \frac{\text{OUTPUT}}{\text{INPUT}},$$

or, $$\text{EFFICIENCY} = \frac{\text{Load}}{\text{Effort}} \times \frac{\text{Distance moved by load}}{\text{Distance moved by effort}},$$

or, $$\text{EFFICIENCY} = \text{Mechanical Advantage} \times \frac{1}{\text{Velocity Ratio}}$$

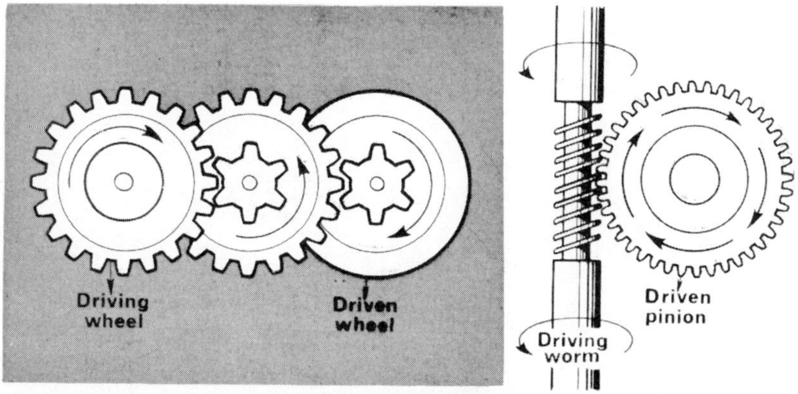

Figure 14.5. On the left is a train of gears as found in the mechanism of a clock. The mechanical advantage of this train of gears is less than 1. On the right is a worm wheel and pinion with a mechanical advantage much greater than 1. In each case the driving gear wheel is on the left of the diagram and the driven gear wheel on the right.

or,   EFFICIENCY $= \dfrac{\text{MECHANICAL ADVANTAGE}}{\text{VELOCITY RATIO}}$

It is only because of the losses due to friction that the mechanical advantage is not as great as the velocity ratio. As the losses are reduced, by lubrication for example, the mechanical advantage increases but it never actually equals the velocity ratio.

Efficiency is normally expressed as a percentage. When we say a machine is 80 per cent efficient we mean that 80 per cent of the work put into the machine is obtained from the machine in its output and the remaining 20 per cent is wasted usually by being converted into heat.

Example. The system of pulleys shown in Figure 14.6 is being used to raise a load of 10 000 N against gravity. The effort needed by the men is only 4 000 N. How efficient is this machine?

By observation it is clear that the load is raised on a pulley block having four ropes supporting it. For every metre that the pulley block rises each rope shortens by one metre and therefore the rope the men are pulling is drawn in four metres. This is another way of saying that the velocity ratio of this machine is 4.

$$\text{The mechanical advantage} = \frac{\text{Load overcome}}{\text{Effort applied}},$$

$$\therefore \text{M.A.} = \frac{10\ 000\ \text{N}}{4\ 000\ \text{N}},$$

$$\text{The efficiency} = \frac{\text{Mechanical advantage}}{\text{Velocity ratio}},$$

$$\therefore \text{Efficiency} = \frac{10\ 000}{4\ 000} \div 4,$$

$$= 0.625.$$

$\therefore$ The efficiency of the machine is 62·5 per cent.

Figure 14.6. This is a pulley-hoist or a block-and-tackle being used to help these men to pull a small boat out of the river. This pulley-hoist has both fixed and movable pulleys. Pulley systems vary, some having more pulleys and some having less.

Questions on chapter 14 (Take g to be 10 m s^{-2}.)

1. Draw a system of pulleys with a single rope passing round two pulleys in each block. If the load is raised by 20 m, how far is the effort moved? What is the velocity ratio of this machine?

2. What is meant by the efficiency of a machine? A man uses a system of pulleys to raise a load of 400 N through a height of 8 m. He has to pull on the rope with a force of 80 N and pulls through 64 m of rope. What is the efficiency of this machine?

3. A pulley system consists of one long rope and two blocks each containing two sheaves. What effort would be needed to lift a load of 200 N if the efficiency of this machine is 35%?

4. A man pulls a body weighing 200 N and mounted on rollers up a smooth inclined plane by means of a rope drawn parallel to the plane. The inclination of the plane is 1 in 16. What is the tension in the rope?

5. Draw a system of pulleys with a velocity ratio of 8. What percentage of the work done is usefully employed if a load of 600 N can be lifted by an effort of 120 N with this machine?

6. Cloth shears used by tailors have short handles and long blades; shears used to cut sheet tin have long handles and short blades. Explain why these shears are made differently.

7. The screw and the wedge are machines that take advantage of the forces of friction. Explain how this is done.

8. Why do racing cycles often have winged nuts instead of ordinary nuts fitted to their wheel hubs?

9. Why is it easier to move a wheelbarrow if the load is placed as near to the wheel as possible?

10. Is there any connection between the development of the screw and how we think the building of the pyramids took place? Draw diagrams to explain your answer.

11. Why is a single fixed pulley used? What is the mechanical advantage of a well-greased perfect single fixed pulley?

12. Define the terms mechanical advantage and efficiency. Draw a diagram of a machine which has a velocity ratio of 4, and describe an experiment to find the efficiency for one particular load. If the efficiency of this machine is 75%, calculate the effort required to raise a load of 120 N.

13. A coal man lifts 5 sacks of coal, each 'weighing' 50 kg, from the ground to 1 m above the ground. How much work does he do? If a loading machine expends 1 500 J of work in carrying out the same job, what is its efficiency?

14. Define mechanical advantage and velocity ratio of a machine. Draw a diagram of a hydraulic system, showing clearly where the effort is applied and where the load force acts. In a hydraulic press, a load of 1 000 N is carried on a piston 0.1 m in diameter. The effort piston is 0.02 m in diameter. What effort will be required to lift the load?

15. The kinetic or molecular theory of matter

There is plenty of evidence that matter is composed of *molecules* in motion. A molecule is the smallest particle of a given substance that can exist on its own. Molecules are built up of still smaller particles known as atoms. For example, a molecule of water contains two atoms of hydrogen and one atom of oxygen. Unfortunately to demonstrate some of this evidence very special and complicated apparatus is needed.

An electron microscope enables us to photograph some of the larger molecules that we cannot see with the ordinary optical miscroscope.

However, we can see the movement caused by the impact of molecules on relatively large particles. We can see this movement, called the *Brownian movement*, in both gases and liquids by means of a low-power microscope. The larger particles appear to be pushed around in an irregular fashion by the impacts due to the smaller molecules.

We cannot see but we can detect with our noses the gradual mixing of gas molecules in air when, for instance, some strong perfume is spread on a handkerchief. The constant agitation by the air molecules causes the dispersal of the molecules of the perfume. This is particularly noticeable if the handkerchief is waved about in the air.

Put a drop of ink carefully on the surface of a glass of water. The ink moves downwards because it is heavier than the water, but it also moves sideways. After an hour or so the ink has spread, like the perfume, all over the space of the container; this is the result of the ink molecules striking not only against themselves but also against the water molecules.

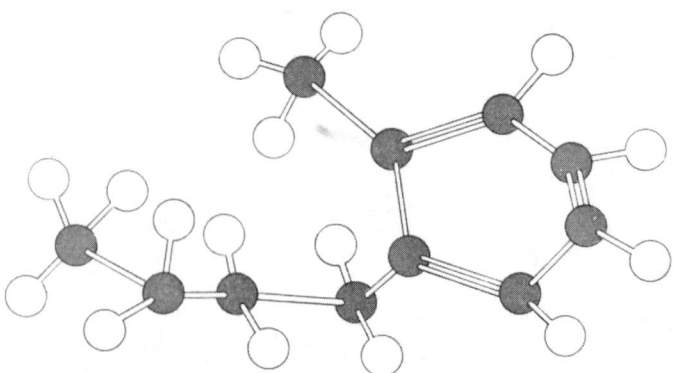

*Figure 15.1.*A complex molecule of one of the many forms of paraffin showing the atoms of which scientists think it is built.

The molecular theory based on this and similar evidence assumes that: (1) molecules are in constant rapid motion, (2) molecules are relatively far apart with huge empty spaces between them, (3) a rise in temperature causes the molecules to increase their speed of movement and, if possible, their distance apart, and (4) molecules exert forces on one another.

Matter, the name used to describe the material of which all bodies are made, occurs in three *states*—solid, liquid, and gas. We usually think of a substance as having only one state: lead as a solid, mercury as a liquid, and oxygen as a gas. However, under the right conditions substances can be made to exist in any one of the three states. Lead can be changed by heating to a liquid and then eventually to a gas. Oxygen can be liquefied and then turned to a solid by cooling it and at the same time applying great pressure to it. Water exists in the solid state as *ice*, in the liquid state as *water*, and in the gaseous state as *steam*.

The molecular theory assumes that: (1) the velocity of molecules is greatest in gases and least in solids, and (2) the molecules in gases have the greatest spaces between them and therefore the greatest freedom of movement; the molecules of liquids have smaller spaces; and the molecules of solids move back and forth along short paths at smaller speeds.

Therefore it is easy to understand why a gas will 'occupy' any space in which it is placed and why it needs an increased pressure to confine it into a smaller volume. Remember the observations made by Robert Boyle (chapter 12). A gas is readily compressed but a liquid and a solid can only be compressed into a smaller volume under very great pressures.

Many of the physical principles and the properties of solids and liquids discussed in earlier chapters, such as elasticity, ductility, malleability, buoyancy, surface tension, cohesion, adhesion, capillary attraction, and the transmission of pressure in liquids can be understood more clearly if we assume molecular forces exist.

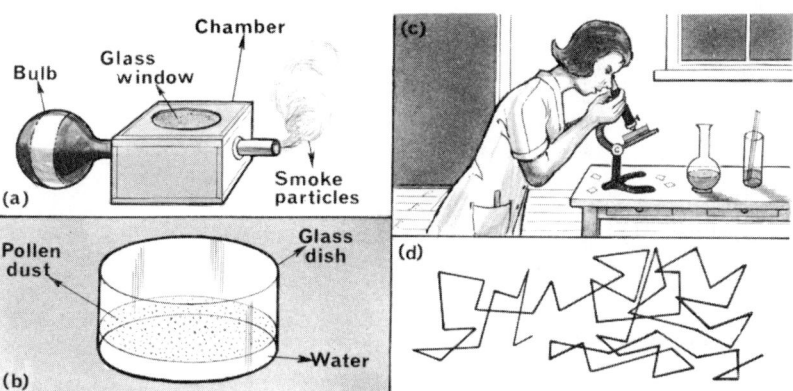

Figure 15.2. The Brownian movement. (*a*) Smoke particles are drawn into the glass-faced chamber for examination. (*b*) Small particles of pollen dust are suspended in water in the glass tank for examination. (*c*) The girl is observing the movement in one of the chambers through a low power microscope. (*d*) This shows the typical haphazard movement of one of the particles.

H

Questions on chapter 15

1. What is the difference between an atom and a molecule? Give some examples to illustrate your answer.

2. In a blown-up football there are an enormous number of molecules and yet the matter present occupies an extremely minute volume of space. Explain this statement.

3. During very cold weather why does a motorist find that the lubricating oil in his engine becomes very sticky?

4. It is believed that all substances consist of molecules. Say how the molecules behave (a) in solids, (b) in liquids, and (c) in gases.

5. Why is it that adhesive tape attaches itself more easily to a smooth surface like glass than to a rough surface like cardboard?

6. What is the Brownian movement and how has it strengthened our belief in the existence of molecules?

7. Why does one sometimes see dust particles moving about in a haphazard way in the air? This is often clearly seen when the air is very still and sunlight shines in through a window into a dull room.

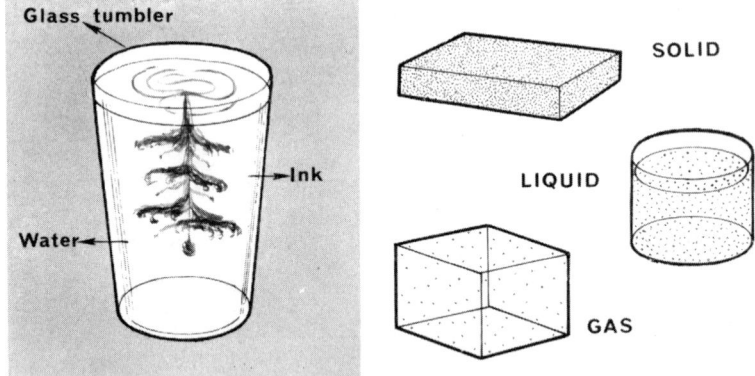

Figure 15.3. This shows how a few drops of ink placed carefully in a glass of water spread downwards and sidewards.

Figure 15.4. The three states in which a substance can exist. The molecules are shown as small dots. Solid state—the substance has a definite shape, its molecules move slowly back and forth in short paths. Liquid state—the substance has to be contained in a solid vessel, its molecules move much faster in longer paths but they pass between one another and do not alter the total volume to any large extent. Gaseous state—the molecules travel rapidly as far as the boundaries of the containing vessel and collide less frequently with one another.

16. Thermal expansion

The heat energy present in a substance is represented by the energy of movement of the molecules that go to form that substance. Heat is a form of energy.

There are many ways of producing heat. When a substance takes in heat, the molecules of the substance move more rapidly and thus have greater energy. We say that the substance gets hotter than it was before it took in the heat.

The human body gets hot as the food consumed is oxidized by the oxygen breathed in; when the brakes of a motor-car are applied the brake shoes rub against the brake drums on the wheels and cause them to become hot—very hot if used on the long descent of a hill; an electric lamp filament gets hot when an electric current is passed through it; the surface of the earth gets hot when the sun shines on it. This last example is of special importance as almost all the energy used by man originally came from the sun. Solar heat energy can be used directly, and solar cells provide the power for the radio transmitters in space craft.

The methods we have considered of producing heat are really methods of transforming one kind of energy into heat energy.

1. Heat from *chemical* processes. The coal or wood of a burning fire combines with the oxygen in the air to liberate heat. This is the chemical process of rapid oxidation. The stored-up energy in food is transformed to heat by the chemical process of slow oxidation in the human body.

2. Heat from *mechanical* energy. The mechanical energy in a moving motor-car is transformed into heat energy in the brake drums when the car is brought

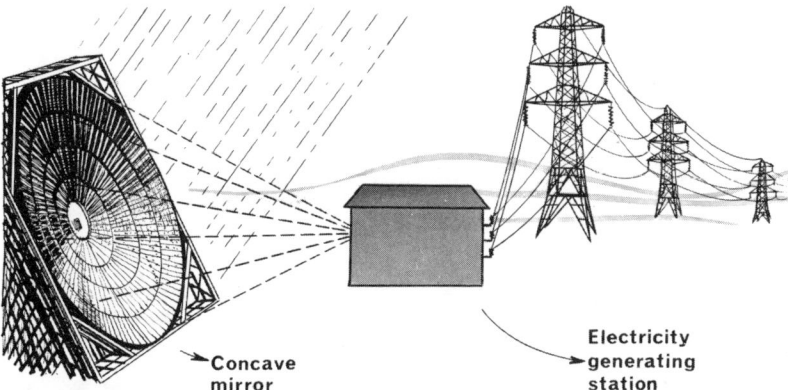

Concave mirror

Electricity generating station

Figure 16.1. A diagram of one type of solar furnace. The heat rays from the sun are directed by the huge concave mirror on to a small area near the focus of the mirror where they are used to heat substances to very high temperatures, or to produce steam so that electricity can be generated.

105

to rest by means of its brakes. The car tyres also become hot.

3. Heat from *electricity*. When electricity passes along the wire of an electric fire the electrical energy is transformed into heat energy.

These three transformations involve energy which was previously created by solar energy and then stored in some form or other. Green plants containing chlorophyll use solar energy in the process of making their foods. In any given time the quantity of energy absorbed by all the chlorophyll in all the plants on earth is more than one thousand million times greater than the quantity of energy generated by the world's biggest power station. After millions of years these plants decay to form coal, natural gas, and oil. Solar energy on the surface of the land and sea causes evaporation and thus lifts water to the skies. The water cools in the skies and forms clouds that in time deposit rain on the hills. The rain finds its way back through waterfalls to the sea. Some of the energy of the falling water is utilized by man for the production of electrical energy.

4. Heat from the *nucleus of the atom*. This is another source of heat which is rapidly becoming important and may one day be the answer to our need for larger and larger supplies of energy. This is obtained by the splitting (*fission*) and the combining (*fusion*) of the nuclei of atoms.

When a substance is heated it becomes hotter, unless it changes its state. The 'hotness' of a substance can be judged very roughly by the sensation it produces as we put our fingers into it or as we touch it. Instruments, called *thermometers*, are used to measure the 'hotness' or *temperature* of a substance with more accuracy than is possible by our sensation of touch.

Suppose we have a source of heat, such as a steadily burning fire or flame, and fill a large can and a small can nearly to the brim with water from the same tap. We can check roughly with our hands that the water in both the cans is at the same temperature. If we then put each in turn for five minutes at the same position above the flames we shall easily observe by touching that the smaller mass of water in the smaller can has risen in temperature more than the greater mass of water in the larger can. Both cans received the same quantity of heat from the source but they register different changes in tempe-

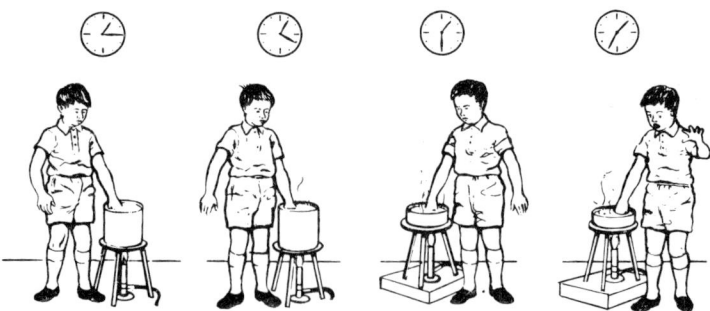

Figure 16.2. An experiment that shows that heat and temperature are different. Two cans on the same source of heat for the same time gain the same quantity of heat but one becomes hotter than the other.

rature. This means that heat and temperature are different. We shall learn later how they are connected with one another, but for the moment we understand that heat is energy and temperature is a measure of 'hotness'. If you put your hand in cold water heat flows from your hand to the cold water and you say that the temperature of the cold water is less than that of your hand. In fact, temperature indicates which way heat will flow when two bodies at different temperatures touch one another. Heat always flows from the body at the higher temperature to the body at the lower temperature.

When a substance is heated and its temperature rises the molecules move faster and cause an increase in size of the substance in all directions. This change is called *expansion* and can, in some circumstances, create enormous forces if one tries to prevent it happening.

Expansion of solids. Solids expand by small amounts, liquids expand by greater amounts, and gases expand considerably more than liquids or solids. When man builds an engine, a machine, a bridge, and many other things too, he must allow for the expansion that may take place as the temperature changes. Our daily lives are very much concerned with the effects of *expansion* and *contraction* of different substances. Any one solid expands or contracts by the same amount for the same rise or fall of temperature, but different solids expand different amounts. The amount of expansion and contraction is very small indeed and usually it is difficult to see unless we make use of some special device for the purpose and observe very carefully. To show the effect the solid is generally made into a long tube or rod and heated along its whole length from the temperature of the air in the laboratory to the temperature of boiling water or steam. The increase in length even then is very small but it is measured by a micrometer screw gauge, or else it is measured on a bigger scale after magnification by a system of levers. Every different metal or solid has its own rate, or coefficient, of expansion. The *coefficient of expansion of a substance* is the fraction of its length which the substance expands when the change of

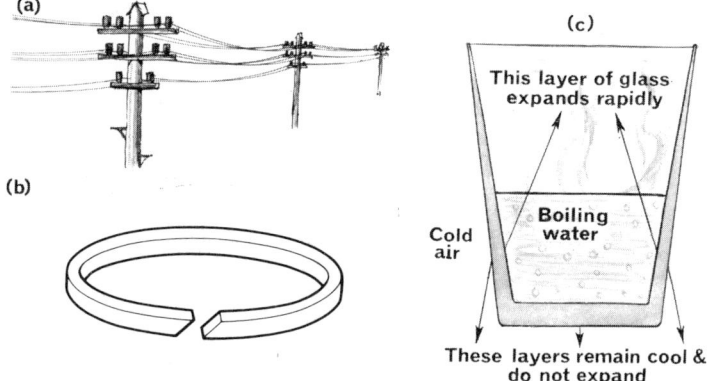

Figure 16.3. Some effects of expansion. (*a*) Sagging telephone wires. (*b*) The gap in a cold motor car piston ring to allow for expansion when the engine is running. (*c*) A thick glass tumbler cracked by boiling water.

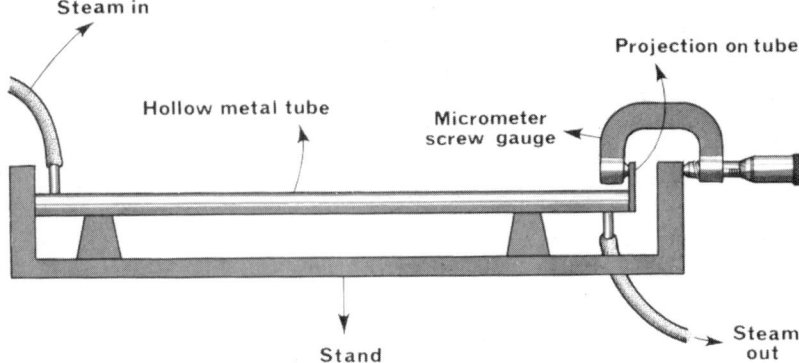

Figure 16.4. A simple method of finding the coefficient of expansion of a metal. Steam is allowed to pass through the tube for some time in order to obtain a steady temperature. The amount of expansion is measured by the change in the setting of the screw gauge.

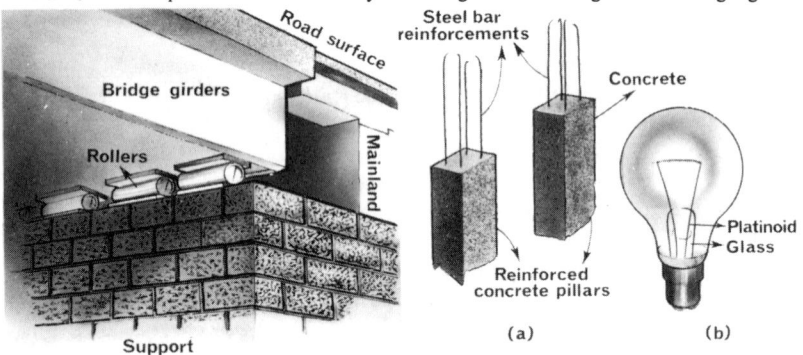

Figure 16.5. Engineers mount bridges on rollers to allow for expansion and contraction. The surface of the bridge interlocks on either side with the mainland surfaces to form smooth connections.

Figure 16.6. The use of two substances possessing the same coefficients of expansion. (*a*) Steel and concrete. (*b*) Glass and platinoid.

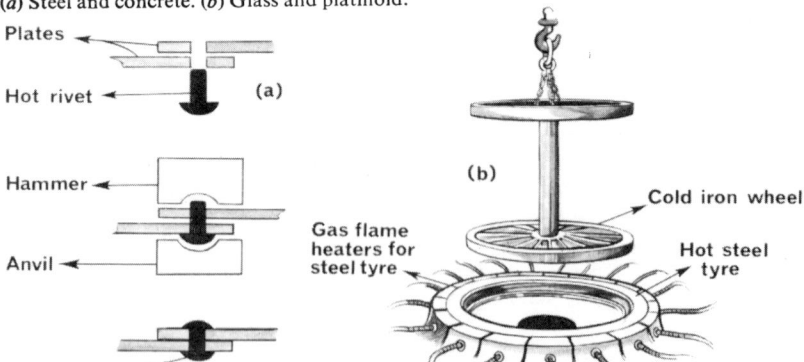

Figure 16.7. How forces of contraction are used (*a*) to seal two metal plates together, and (*b*) to secure a steel tyre on an iron wheel.

temperature is one degree centigrade. The coefficient of expansion of a substance is a very useful quantity, for it enables us to calculate in advance what the actual expansion of any length of that substance will be when heated any number of degrees. For example, when engineers build a bridge they can calculate exactly how much the bridge will expand on a hot day.

Example 1. Calculate the coefficient of expansion of iron given that a hollow iron tube 100 cm long at 17° C is heated by passing steam through it until the whole tube is at 100° C, and that during this heating the expansion is measured to be 0.10 cm.

	100–17° C		100 cm		0.10 cm
An increase		causes		long	
of	83° C	a	1 cm	to	0.001 cm
temperature of		tube		expand	
	1° C		1 cm		$\dfrac{0.001}{83}$ cm

or 0.000 012 cm

∴ The coefficient of expansion is 0.000 012.

The coefficients of expansion per degree centigrade of a few common substances are given in the following table:

Aluminium	0.000 023	Invar	0.000 000 87
Brass	0.000 018	Copper	0.000 017
Steel	0.000 011	Glass	0.000 009
Concrete	0.000 011	Platinum	0.000 009

You will notice that *invar*, an alloy of steel and nickel, has a very small coefficient of expansion indeed and that is why it is now used for making the pendulums of accurate clocks. Steel and concrete have equal coefficients of expansion. This is fortunate for it enables us to construct concrete girders, beams, and bridges, reinforced with steel rods without fear of them breaking as temperature changes occur. Platinum and glass have also equal coefficients of expansion and this explains why we can seal platinum wires in glass without fear that the glass will crack as it is heated or cooled. The wires carrying the electric current into an electric lamp were made of platinum where they pass through the glass bulb. This property of platinum and glass made possible the construction of an electric lamp. A much cheaper alloy, platinoid, with the same coefficient of expansion as that of glass is now used instead of pure platinum because it is much cheaper.

Fortunately we can find many good uses for the enormous forces that occur during expansion and contraction. These forces are often used to tighten up some things or to push others. We can mention only a few of the uses here but you will find many more as you look around.

The plates of a boiler can be joined firmly together by *rivets*. A red hot rivet is inserted through two holes opposite one another in two plates, hammered flat, and then left to cool. As the rivet cools it contracts and pulls the plates against one another.

Steel *tyres* for locomotive wheels are made so that they are just too small to fit when cold. When the tyre is heated all round it expands, and then the cold wheel is lowered into it. At this moment the tyre just fits around the wheel but when it is cooled it grips the wheel securely.

A glass *stopper* stuck tight in a glass bottle can be loosened by heating the neck of the bottle either with a small flame or by rubbing it rapidly with a cord. In the same way a tight metal cap screwed on a glass bottle can be heated so that it expands and is made easier to remove.

A steel *liner* is inserted into an iron cylinder block of a petrol engine by cooling the liner in a refrigerator and then tapping it into position when cold. As the liner warms, it expands and becomes firmly held inside the iron cylinder block.

Iron and brass have different coefficients of expansion—that of brass is nearly 1½ times that of iron. Thin long bars of these two metals are riveted together along their entire length to form a *bimetallic strip*. An increase in temperature causes a bimetallic strip to become curved because of the unequal expansions of the two metals. This bending is used in many ways—to operate thermostats, to record temperatures, to ring fire alarms, and to compensate for temperature changes in the balance wheel of a watch so as to keep the period of vibration constant.

Unfortunately there are cases where the expansion and contraction of substances cause us difficulties and even danger, so we have to devise ways of overcoming them.

Railway lines used to be laid in short lengths jointed together by plates in such a way that gaps were left between them to allow for expansion and contraction as the temperature varied. These gaps were responsible for the clickety-click noise as the train bumped over them. But it has been found that when these lengths of rail are welded together to form one continuous rail and held securely by strong sleepers well embedded in ballast, expansion is only possible at the ends. The force of expansion in the centre of the section of rail simply causes internal forces which cannot even alter its shape. Sliding expansion joints (Figure 16.8) compensate for the expansion at the ends of the continuous rail.

Figure 16.8. The expansion joints at the end of very long continuous welded rails. Note how the rails slide by the side of one another and how the sleepers at the joint are held together.

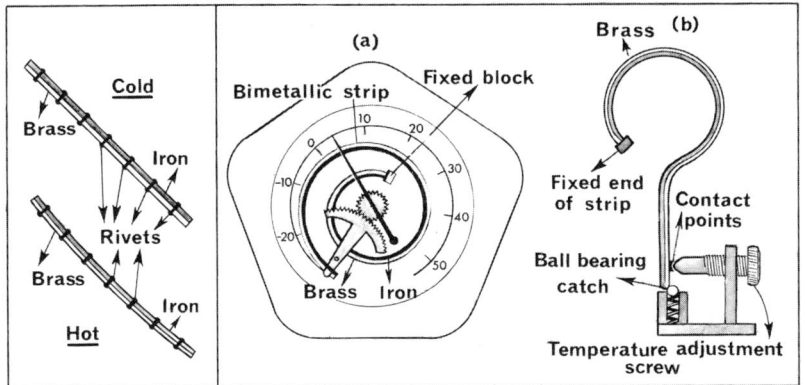

Figure 16.9. The bending of a bimetallic strip due to a change in its temperature.

Figure 16.10. A bimetallic strip used in (*a*) a thermometer, and (*b*) an electric heater thermostat where it cuts off the heater current at a pre-determined temperature by clicking over the ball-bearing catch, and reconnects it when the temperature falls.

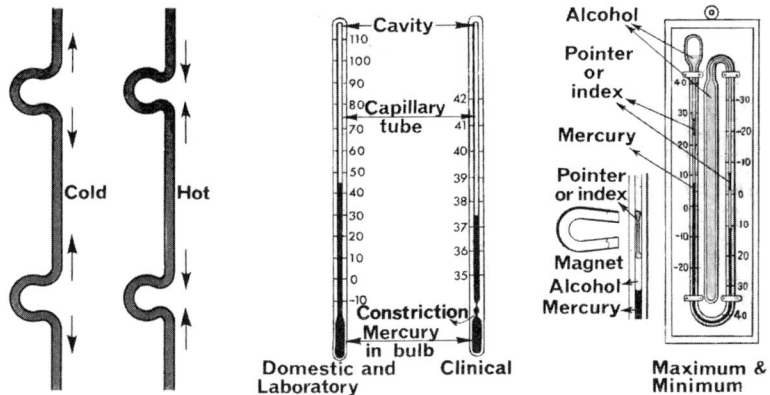

Figure 16.11. The same long pipe carrying a hot liquid (right) and a cold liquid (left). Notice that the ends of the pipe are fixed and how the expansion loops take up the movements due to expansion and contraction.

Figure 16.12. Some thermometers in common use. (Not drawn to scale.)

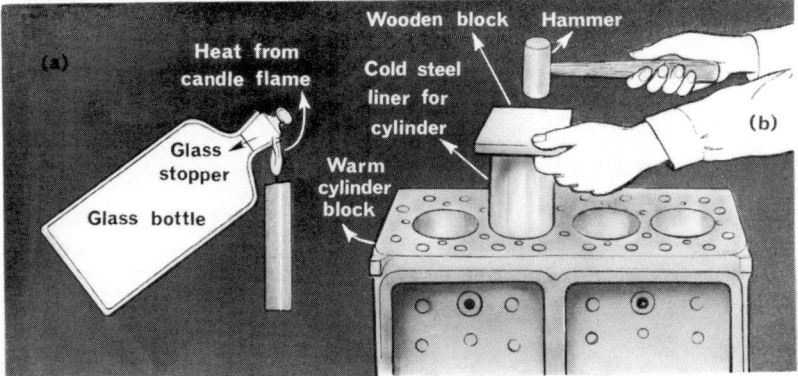

Figure 16.13. How forces of expansion are used (*a*) to release a tight glass stopper from a bottle, and (*b*) to fix a steel liner inside an iron cylinder block.

When a motor-car engine starts, the temperature of the pistons rises more than that of the cylinders. Thus the expansions of piston and cylinder are unequal. That is why, when cold, the pistons are slightly smaller than the cylinders in which they move. When hot, the pistons fit well but not too tightly. *Piston rings*, fitted round the piston, prevent the gases escaping by sealing up the space between the piston and the cylinder wall when they are both cold. The piston rings are themselves split to allow for expansion.

Long pipes that carry steam or any hot liquids or gases have *flexible expansion loops* built in them at regular intervals so that any expansion or contraction may be absorbed.

The expansion of liquids is used in a thermometer to indicate the temperature. The expansion of mercury is almost uniform throughout the whole range of temperatures that we normally wish to measure so that a mercury thermometer has equally spaced scale divisions. Alcohol has a uniform expansion at lower temperatures.

Liquid thermometers are made in a variety of ways for various purposes but each has a large bulb at the end of a long tube. The liquid fills the bulb and part of the very fine bore in the long narrow tube. As the temperature rises, more of the liquid from the bulb expands along the bore. The number of the mark to which the liquid reaches on the scale tells us the temperature of the bulb.

Celsius, a Swedish astronomer, designed the centigrade scale of temperatures by dividing the interval between the temperature of the melting point of ice and the temperature of the boiling point of water at a pressure of 76 cm of mercury into 100 divisions. These temperatures have become known as the *fixed points* of a thermometer scale.

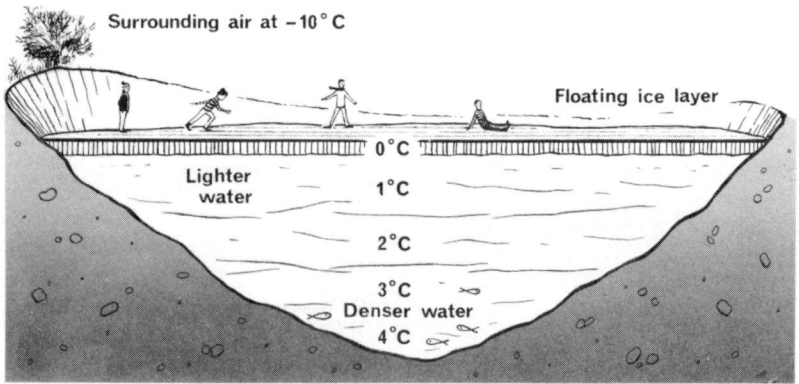

Figure 16.14. This diagram of the cross-section of a lake shows the temperatures that are likely to be found on a cold day in winter when the air temperature is lower than the freezing point of water.

A certain temperature on the centigrade scale of temperatures in many countries is spoken of as so many degrees celsius and written °C, but in Britain it is still customary to speak of degrees centigrade and write the symbol for this as °C.

The human blood temperature is very nearly 37° C.

Mercury thermometers have many uses in the home. When you are making toffee the sugar mixture is stirred and heated to a temperature of 121° C precisely. If you wish to liquefy some granulated honey, you must slowly warm it in water and then keep it for some time within the limits of 62° C to 64° C: it does not liquefy at a lower temperature, and at a higher temperature it loses some of its fine qualities. When you prepare the water to bath a baby, the water should be heated to 37° C. All these temperatures are conveniently read on a mercury thermometer.

The *expansion and the contraction of water* is unique. Water starting at 0° C contracts as it is heated until 4° C is reached, after which it expands as the temperature rises. This means that every given mass of water has its smallest volume and greatest density at 4° C. We do not know why this is so, but because this happens the water at the bottom of a lake in winter is usually at 4° C even when there is the lighter layer of ice on the top of the lake. In this way the aquatic life at the bottom of lakes survives without difficulty during a severe winter.

Expansion of gases. All gases expand when heated by about the same proportion, and this is almost 20 times that of mercury or 8 times that of water. It was a French physicist, called Charles, who first observed that *all* gases have very nearly the same coefficient of expansion. Charles's law says that if a gas is kept at a constant pressure then it will expand by $\frac{1}{273}$ of its volume at 0° C for every rise in temperature of 1° C. This led to the guess that if a gas could be cooled through 273° C starting at 0° C then its volume would be reduced by $\frac{1}{273}$ for each 1° C until it reached zero volume. It is at this temperature of −273° C that the molecules would not move or hit the walls of the containing vessel. Nobody has so far been able to cool a gas to −273° C or, as it has been called the *absolute zero* of temperature.

We can use Charles's law to find the relationship between the volume of a gas and its temperature on the kelvin scale based on this fundamental zero. The kelvin temperature is 273 plus the celsius temperature.

Charles's law can be stated in the form of a simple equation:

$$\frac{\text{Original volume}}{\text{New volume}} = \frac{\text{Original absolute temperature}}{\text{New absolute temperature}}$$

or

$$\frac{V_1}{V_2} = \frac{T_1\,K}{T_2\,K}$$

Example 2. In the morning a small light balloon was partially filled with 1 000 cubic metres of helium gas at 15° C. Later in the day the atmospheric pressure was the same as it was in the morning and the fabric of the balloon weighed the same, but the balloon was observed to have swollen. What was its new volume if the temperature had risen to 27° C?

Original volume = 1 000 m³, original temperature = (273 + 15) K, new temperature = (273 + 27) K and the new volume = V_2. Using Charles's law:

$$\frac{V_1}{V_2} = \frac{T_1 \text{ K}}{T_2 \text{ K}}$$

$$\therefore \frac{1,000}{V_2} = \frac{288}{300},$$

$$\therefore V_2 = 1\ 042$$

∴ The new volume of the balloon was 1 042 cubic metres.

Flour is mixed with water, salt, and yeast to make bread. The mixture, called dough, is left to 'rise' for some hours. The yeast plant produces carbon dioxide gas in little pockets throughout the dough. When this 'risen dough' is baked in the oven the carbon dioxide and the steam from the water expand, as all gases do, to produce a large number of holes in the bread.

It was Galileo, an Italian astronomer, who invented an *air thermometer* as well as many other scientific things. His air thermometer is extremely sensitive to changes in temperature and is very easily made. Unfortunately it is also sensitive to changes of atmospheric pressure otherwise it would be used more frequently.

If a certain volume of a gas is heated it becomes lighter, and if it is surrounded by cooler denser gas and free to move it will then rise. This movement forms *convection currents* if continued long enough. If the gas is air and is enclosed

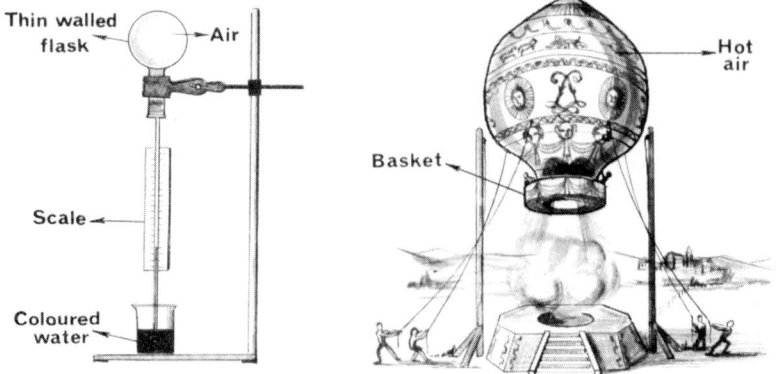

Figure 16.15. An air thermometer invented by Galileo. The liquid in the tube and reservoir is coloured water.

Figure 16.16. A hot-air balloon was used when man first conquered the air. A diagram of the balloon used for this conquest by the Montgolfier brothers in France in the year 1783.

in a balloon it will raise the balloon. The balloon will continue to rise until the weight of the balloon and its gondola is equal to the difference between the weight of the cold air that has been pushed aside by the balloon and the weight of the hot air in the balloon itself.

Low temperatures. Although a gas has not been cooled to the absolute zero of temperature ($-273\,°C$ or $0\ K$), temperatures very close to this have been reached in the Clarendon Laboratory at Oxford and some peculiar effects have been observed.

Most real gases liquefy or solidify if they are cooled. At $-78\,°C$ or $195\ K$ carbon dioxide forms the solid known as 'dry ice'. Pucks whose interior is filled with 'dry ice' are used in schools on large flat glass sheets to observe the movements of bodies. The 'dry ice' is slowly melted back to carbon dioxide which escapes below the puck and forms a cushion of gas on which the puck moves about almost without any friction. The same principle is employed by hovercraft vessels. Considerable use of this solid is made in preserving ice-cream and frozen foods, in transporting by rail or road containers of fresh fruit and vegetables, and in making fizzy drinks such as mineral waters, beer and cider.

At $-160\,°C$ or $113\ K$ liquefied methane, which is the main ingredient of natural gas, can be stored in large holders below ground level. This is interesting because the holder utilizes frozen ground for its casing, so that all one has to do is to excavate a large underground hole. The low temperature of the liquefied methane freezes the earth until it resembles a hard concrete basin which is no longer porous.

At $-183\,°C$ or $90\ K$ oxygen forms a pale light blue liquid and in this state can be used as a fuel in space rockets. Rubber, fruit and flowers placed in the liquid become hard and brittle at this temperature and are thus easily powdered.

The electrical resistance (see chapter 27) of most metals decreases as the temperature is lowered. However, some metals, for example tin and lead, and some compounds have a most extraordinary property, for their resistance falls suddenly to zero at a particular low temperature. These are then called *superconductors* through which, once started, a current will go on flowing without any cell, battery or other means to maintain it.

Questions on chapter 16

1. Define the term coefficient of linear expansion. Describe a simple method for determining this quantity for brass. Describe what happens to its volume when some ice at $-5°$ C is heated until the temperature of the water formed is $10°$ C.

2. A metal rod is 100 cm long at $0°$ C, and 100.2 cm long at $100°$ C. What is the coefficient of linear expansion of the metal?

3. Describe an experiment that will show that the increase of a certain volume of a gas at constant pressure is considerably greater than the increase in the same volume of a liquid for the same rise in temperature.

What precautions need to be taken to overcome the effects of expansion in (a) a railway line, and (b) the pendulum of a clock?

4. Draw a diagram of a clinical thermometer. State two important ways in which this thermometer differs from an ordinary thermometer.

5. In what ways is a clinical thermometer different from an ordinary one? Why should it never be sterilized in boiling water?

6. Explain with the help of a diagram, how you would check that the $0°$ C mark on a thermometer was correct.

7. Explain why there is a lowest possible temperature but no highest possible temperature.

8. Two identical vacuum flasks are partially filled, one with hot water and the other with cold water. They are kept for several hours in a room at a temperature in between the temperatures of the hot and cold water. Which stopper is then found to fit more loosely and which to fit more tightly. Why?

9. Why can 'Pyrex' glass withstand large and rapid changes of temperature?

10. When the mercury thread of a thermometer is observed carefully it is found that as the thermometer bulb is dipped into hot water the thread drops a little before it begins to rise. Why is this?

11. A steel bolt has become stuck inside a brass bush. How can it be released and why does the method you describe actually work?

12. The water cooling system of a motor-car has just boiled dry. Why is it unwise to refill immediately with cold water?

13. A motorist fills with cold water the water cooling system of his car when it is cold. Why will some water overflow as the engine heats when it is running?

14. When green wood is put on an open fire why does it crackle and shoot off sparks around the fire?

15. Why do fountain pens that are nearly empty often leak ink when being used?

16. Why do small chips often break or fly off large rocks on hot sunny days?

17. Why are concrete roads laid down in sections with bitumen poured in between the sections?

18. Why is a knowledge of the coefficient of expansion of certain materials necessary in the following cases: (a) for a dentist when filling decayed teeth, (b) for an engineer when making pistons for internal combustion engines, and (c) for the manufacturer of enamel for bathtubs?

19. When we make flame tests of salts in chemical detection we use a platinum wire fused into a glass rod to carry a small quantity of the salt. Why does the platinum wire not splinter the glass rod or fall out of it when it is heated to red heat?

17. The transfer of heat

Heat can be transferred from one point to another in three different ways: *conduction*, *convection*, and *radiation*. Frequently a body gains or loses heat by a combination of all three ways.

Conduction takes place when heat is transferred from particle to particle throughout the whole of the body without any visible signs of movement at all. The kinetic theory suggests, however, that the movement of the molecules is passed from one molecule to the next along the body when one part of the body is made hotter than another. The molecules of the heated part having gained a high speed bump their slower moving neighbours in the cooler part. But this movement of the molecules is not visible. Some substances, particularly metals, are good heat *conductors*, whilst others are so poor that we can use them as *insulators* to hinder the flow of heat. Copper, aluminium, and iron are good conductors. Asbestos, glass fibre, sawdust, wood, fur, wool, and air are poor conductors.

Does this explain why saucepans have copper or aluminium bases? Why heating stoves are made of iron? Why soldering irons have copper tips and wooden handles? Why attic floors in houses are often covered with a layer of glass fibre? Why the fur on animals grows thick in winter, and man wears woollen clothes in the cooler climates? There are many other uses of good conductors and good insulators that we can learn about if we are really observant.

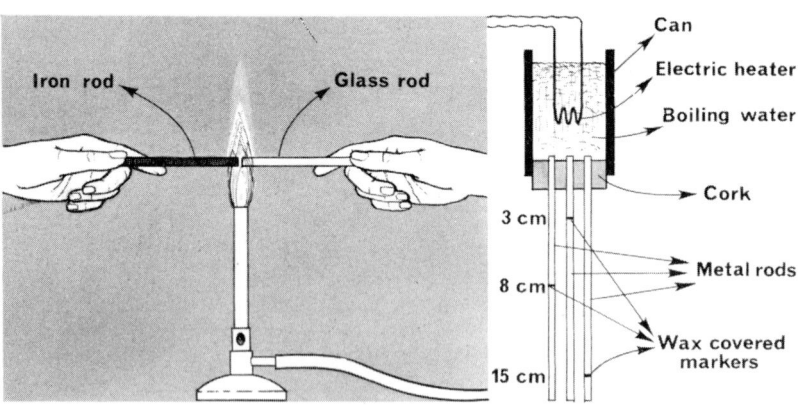

Figure 17.1. An experiment to find out which conducts heat more readily—an iron rod or a glass one. If these two rods are held in the same way in a flame it soon becomes evident which is the better conductor, iron or glass.

Figure 17.2. The experiment designed to compare the heat conductivity of metals. The small electric immersion heater keeps the water boiling in the can. The distances from the ends of the rods to the markers are shown.

It is possible to find out by a simple experiment how good a metal is as a conductor, i.e. its *conductivity*. Rods of various metals having the same diameter are inserted through holes in the cork that forms the base of a can. Small wax-covered markers are attached to the rods at the top just below the cork. Boiling water is poured into the can and heat from the water begins to be conducted down the rods. The markers start to fall as the heat melts the wax that holds them on the rods. After some time the markers do not fall any further and the position they reach tells which rod is the best conductor and which the worst. A typical experiment with three common metals to find the comparative values of their heat conductivity would be:

Copper	92
Aluminium	48
Iron	14

Liquids are in general poor conductors of heat. It can be shown that water is a poor conductor of heat by placing at the bottom of a thin-walled test-tube a small piece of ice. Slide a coil of solder wire down the tube to the ice, so as to keep the ice in position. Nearly fill the test-tube with cold water. Heat the water near the top of the test-tube with a small flame. The fact that you can boil the water there for a very long time without the ice melting indicates that little heat travels down the water in the test-tube. Therefore you conclude from this experiment that water is a poor conductor of heat.

Gases have a much lower conductivity than liquids. Substances that enclose small volumes of air, like certain makes of underwear and well-teased wool, are very poor conductors of heat. That is why wool and certain types of man-made fibres are made into blankets and warm clothing. Light powdery snow that has fallen gently to the ground contains pockets of air and thus it acts like a blanket and prevents heat escaping. In this way it protects the roots of fruit trees, winter wheat, and other young crops from the extreme cold of the winter weather.

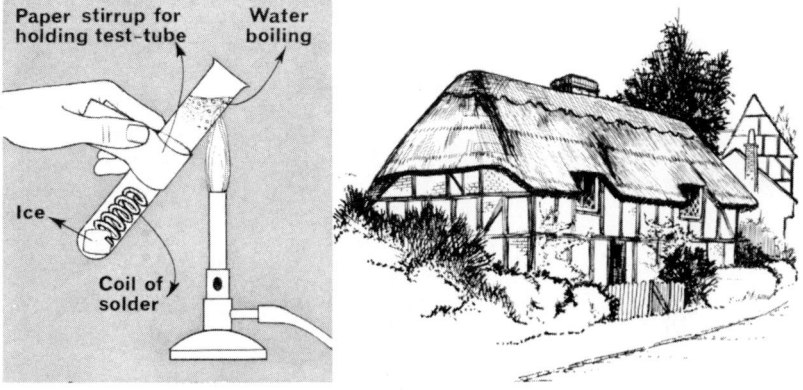

Paper stirrup for holding test-tube

Water boiling

Ice

Coil of solder

Figure 17.3. An experiment to show that water is a poor conductor of heat.

Figure 17.4. This house is warm in winter and cool in summer because the thatch on its roof has a low conductivity of heat due to the countless small pockets of air it encloses.

I

Convection is the process of conveying heat by the circulation of heated portions of a liquid or a gas. Heat causes a fluid (liquid or gas) to increase in size and thus to become less dense. The cooler and denser fluid pushes the warmer and lighter fluid upwards. These movements are called *convection currents*. The winds are convection currents caused by the heat of the sun. Everyone has seen the swift upward movement of the hot air above a large bonfire and how far the glowing sparks will travel. We learned in the last chapter that hot air balloons rise for the same reason.

Similar convection currents are useful because they can cause a strong draught of air through a furnace, warm a building with hot air, and ventilate coal mines. Glider pilots rely on 'thermals' for lift. If you are observant you will find many other purposes for which convection currents are used.

A simple paper spiral cut from a single sheet of paper and hung up on a length of cotton will enable us to detect the presence of convection currents in air. The spinning of this spiral will indicate the position of rising hot air or falling cold air.

A hot water storage tank is often heated by an electric immersion heater. The water is stored in a tank surrounded by a good insulator and then a protective casing. Convection currents of water circulate throughout the tank until all the water reaches the temperature at which the thermostat cuts off the electricity. Hot water pours from the delivery tube at the top of the tank when cold water enters by the tap at the bottom.

Radiation is another method of heat transfer. Some solid, liquid, or gas has to be present for the transfer of heat by the two ways so far studied—conduction and convection. But a body can receive and give off heat through an empty space. For example, we can feel the warming effect of the sun's rays although we know there is nothing at all in the space between the sun and the earth. This transfer of heat is by radiation. We think radiant heat travels in waves in the same way as light travels. Radiation can also travel through space

Figure 17.5. Upward hot air convection currents above a bonfire.

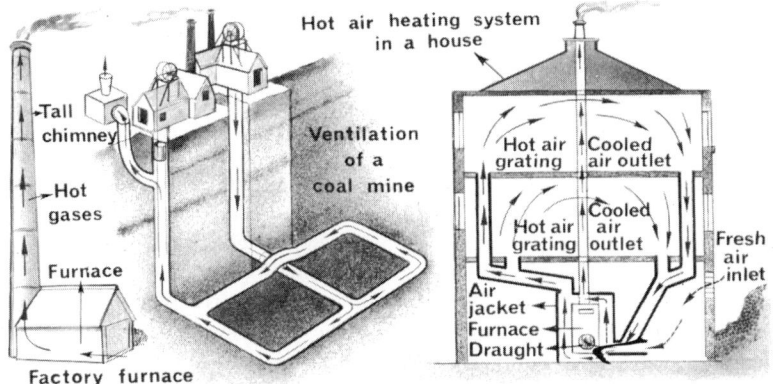

Figure 17.6. How convection currents in air can be used to help man both in his work and in his home.

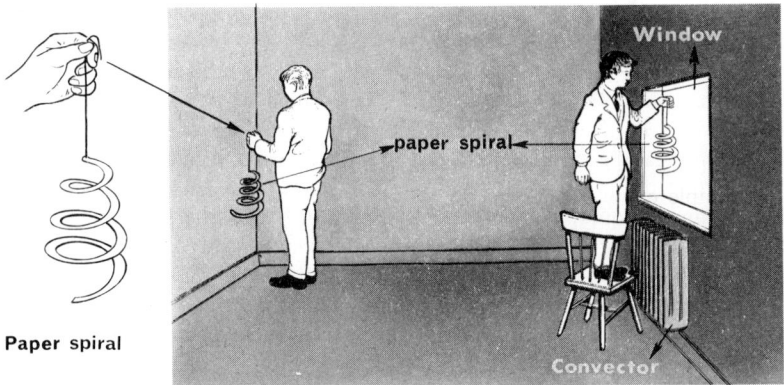

Figure 17.7. A detector of convection currents in air.

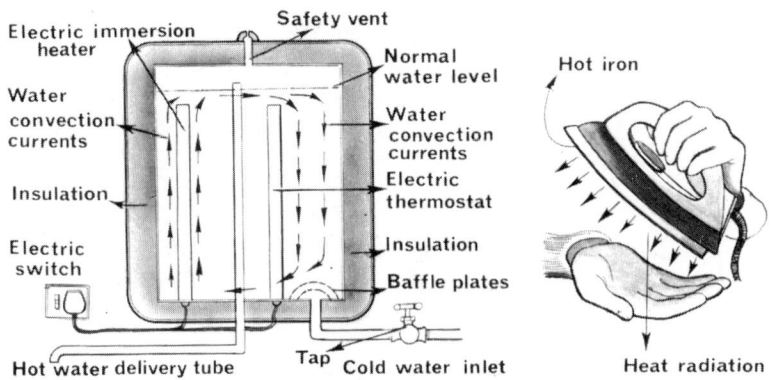

Figure 17.8. A hot water storage tank heated by electricity.

Figure 17.9. The transfer of heat by radiation through the air from an electric iron.

filled with air, and in this case the molecules of air play no part in the transfer.

When heat radiation strikes a body three things can happen. It can be reflected, transmitted, or absorbed according to the nature of the body and its surface. In each case it behaves in the same way that light behaves and obeys the same laws of reflection.

Absorption of heat radiation depends on the nature of the surface of the absorbing body. Dark matt surfaces are good *absorbers* and poor *reflectors*. A black object strongly absorbs heat radiation. Shiny light-coloured surfaces are poor absorbers and good reflectors of heat radiation. Bodies whose surfaces are good absorbers are also good radiators, and similarly poor absorbers are poor radiators. These properties concerning the absorbing and reflecting powers of surfaces will enable us to answer questions like these. Why do tea-pots and hot-water jugs have shiny surfaces? Why is the motor-car radiator (the real one inside the bonnet) painted black? Why do men wear light-coloured suits in the tropics? Why does a piece of dark metal or a leaf sink quickly into snow or ice when the sun shines upon it? Why do fruit growers take precautions against frosts on cloudless nights in spring?

A simple experiment to show that dull black surfaces absorb heat more easily than shiny white surfaces can be performed with the apparatus shown in Figure 17.11. The two copper sheets of the same size are coated, one with soot from a candle flame, and the other with glossy white paint. A little paraffin wax is melted and spread over the reverse side of each sheet. The sheets are then mounted equidistant from some source of heat, either a candle flame or an electric heating coil. By observing the effect on the wax of the heat absorbed the better absorber can be determined.

There are many ways of heating a room in a building. The one we shall consider involves all three ways of transferring heat. The most important is convection because it causes the circulation of hot water in the pipes and of hot air in the rooms. Radiation and conduction are involved in heating the

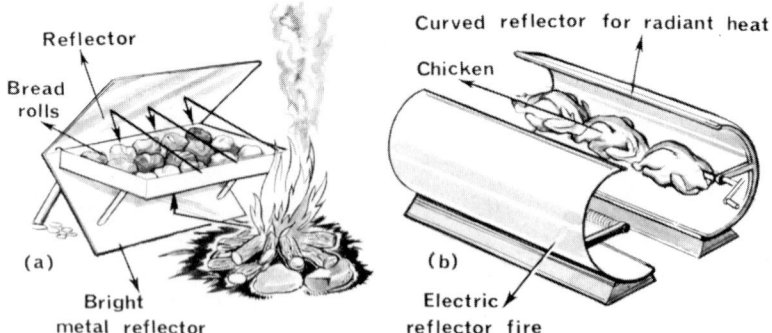

Figure 17.10. Cooking by radiant heat. (*a*) A simple method of cooking small bread rolls by the radiant heat from a camp fire. (*b*) A chicken barbecue—rays of heat strike the revolving chicken from all sides—by direct radiation in front and by reflection from the curved mirror behind.

water in the furnace and the air in the rooms. Can you understand how conduction and radiation are involved in these processes?

The *convector* in the room also radiates heat to a small extent and is for this
reason sometimes called a *radiator*. This is not a good name for it as it only
radiates a very small quantity of heat whereas it convects much more. Heat is
conducted from the hot water inside the convector through the iron of which
it is made to the air in contact with the outside surface of the convector. The
air is heated in this way and then begins to circulate by convection currents.
Sometimes when convectors are close to walls the dust carried by rising hot air
currents makes dirty marks on the surface of the walls above them.

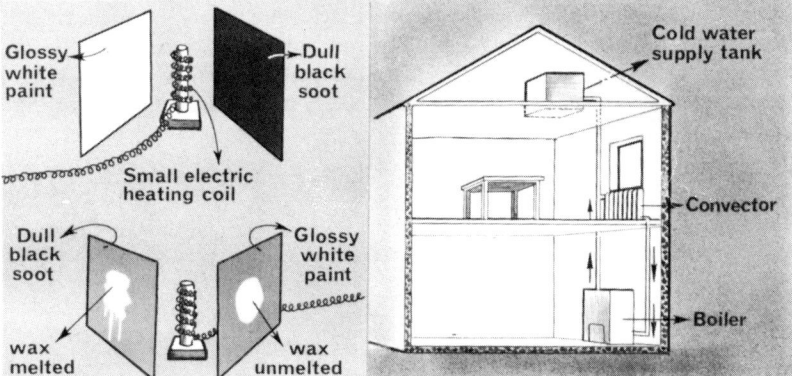

Figure 17.11. The apparatus used to compare the absorption powers of two different
surfaces seen from both sides.

Figure 17.12. Convection currents in the water pipes carry heat from the boiler to the
convector in the room above.

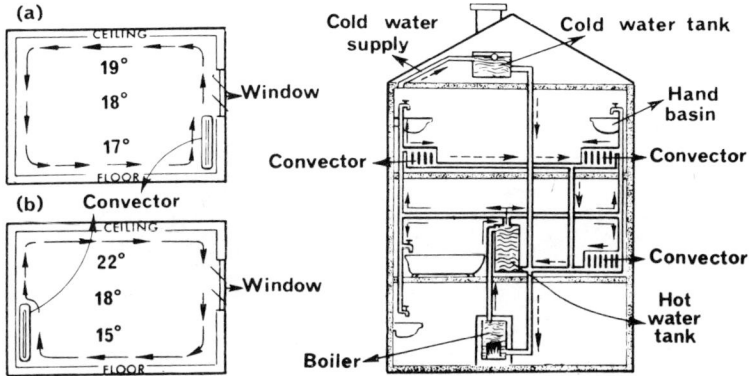

Figure 17.13. Air convection currents set up by convectors in two rooms. (*a*) The cold
fresh air entering by the window is heated immediately and circulates giving warm fresh
air throughout. (*b*) The cold fresh air travels across the floor of the room cooling the
feet of everybody in the room before it reaches the convector and is heated. The figures are
possible air temperatures measured in degrees centigrade.

Figure 17.14. A hot water heating system in a house designed to supply convectors, hand
basins, and bath. Why is it desirable to place the hot water tank above and close to the
boiler?

It is possible to use the same boiler in a house to heat the water used for warming the rooms and also to supply the hot water tap of the hand basin and bath.

There are several types of true radiators of radiant heat used for heating purposes in homes and buildings. They may be heated by hot water, steam, or electricity, and may be installed in the walls or on the ceilings of rooms. The essential condition is that the temperature of the surface emitting the radiant heat should be as high as possible. If this condition is fulfilled they give out more heat by radiation than by convection. Why are radiators mounted in the ceiling unlikely to cause convection currents at all?

Another type of radiator is the electric reflector fire that is simple, light, and convenient. It can be adjusted to send its beam of radiant heat in any desired direction. The electric heating bar is placed at the focus of the parabolic mirror so that the heat rays after reflection are nearly all radiated in the form of a parallel beam.

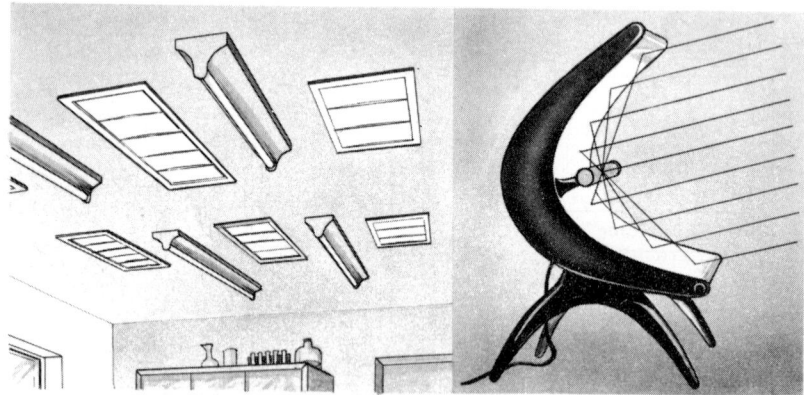

Figure 17.15. Radiators installed in the ceiling of a school laboratory. Even on cold days the windows are left open but the pupils feel warm because of the radiation. The radiators are painted with a matt cream surface. Why do you think they are not painted matt black like the radiator of a motor-car?

Figure 17.16. A modern portable electric reflector fire. This is a true radiator. The thin lines represent the paths of the radiant heat rays from the electric heating bar that is placed at the focus of the parabolic reflector.

Questions on chapter 17

1. Explain how heat is transmitted from place to place by (a) conduction, (b) convection, and (c) radiation. Show clearly in what ways these three methods of heat transfer differ.

Describe and explain how sea breezes are formed on a hot day near the coast.

2. What fact is true about one method of heat transfer that does not apply to the other two?

3. Why do you not burn your fingers while holding a burning match?

4. Why are hollow bricks often used in building construction?

5. Why is a coil of wire often used for the handle of a poker?

6. Why can Eskimos keep warm in ice igloos?

7. Why is it possible to use sheets of newspaper instead of a pullover to keep you warm?

8. Why do double windows help to keep a house warm in winter?

9. Wood is a better conductor of heat than air. Why, therefore, is sawdust used as an insulator in some refrigerators and ice storage houses?

10. Why is it dangerous in extremely cold climates to touch metal objects with bare hands?

11. Why does a thick aluminium saucepan feel colder than a wooden table when both have been in the same room for a long time?

12. Explain why (a) saucepans often have non-metallic handles, and (b) fluffy materials keep you warmer than smooth ones of the same weight.

13. Why can you crawl across the floor of a smoke-filled room when you would be choked if you stood upright?

14. Why is it often cooler in valleys than on neighbouring hillsides on a still summer night?

15. Why is a tall chimney built for a factory boiler which uses natural convection currents to provide the air draught for its fire?

16. Make a sketch of a motor-car engine and its radiator and mark on it with arrows the directions in which the convection currents should flow.

17. Explain why smoke rises up a chimney. Is it pushed up or sucked up?

18. Why does a milk bottle of thick glass crack if boiling water is poured into it when it is cold?

19. Explain why (a) electric fires frequently have highly polished metal behind the heating elements, and (b) hot water radiators are usually placed low down in a room.

20. Sheets of aluminium foil are often used as insulation, expecially in the roofs of houses, and also as a precaution against burning food when cooking it in the oven. Explain how this is possible in spite of the fact that aluminium is a good conductor of heat.

21. Why does snow melt more rapidly on a dirty well-trodden footpath than on the ground near the path?

22. Explain why you may feel the heat rays from the sun when sitting behind a glass window in the winter although the window itself is very cold.

23. We are informed that the temperature on the surface of the moon in sunlight is over 100°C and during the periods of darkness is well below 0°C. Why is the range of temperatures so much greater than that on the earth?

24. Sometimes leaves fall on a snow surface. Why after several days of sunshine have they sunk down into the snow?

25. Why is ground frost unlikely on a cloudy night?

26. A kettle of water is placed on wood fire. Explain how all three processes of transference of heat are involved in bringing the water in the kettle to the boiling point.

27. A vacuum flask keeps hot liquids hot and cold liquids cold for a long time. Explain how all three processes of transference of heat are minimized by a well made vacuum flask.

18. Change of state

We are now able to understand the heat changes that occur as a substance turns from a very cold solid, to a liquid, and then to a hot gas. Let us consider a fixed quantity of very cold ice at $-20°$ C. The molecules of ice move slowly back and forth along short paths hitting and bouncing off one another.

If we give these molecules some heat energy they agitate more rapidly and bounce apart more frequently until the temperature of the ice rises to its melting point at $0°$ C. The heat energy supplied is changed into the energy of the moving molecules. Each kilogramme of ice requires a certain quantity of heat energy measured in joules to raise its temperature by one degree celsius. This quantity of energy is called the *specific heat capacity of ice*.

In the same way the quantity of energy, also measured in joules, required to raise one kilogramme of water one degree celsius is called the *specific heat capacity of water*.

The word *specific* simply means *per unit mass*. Most substances have a smaller specific heat capacity than water, and some are listed below.

Specific heat capacities measured in joules per kilogramme per degree celsius

Lead	135	Glass	670	Meths	2 500
Copper	400	Earth	840	Sea water	3 900
Iron	460	Ice	2 100	Water	4 200

When our fixed quantity of ice is heated sufficiently to reach its melting point the ice gradually changes to water without any further rise in temperature. We continue to supply heat energy at a steady rate to the melting ice. It

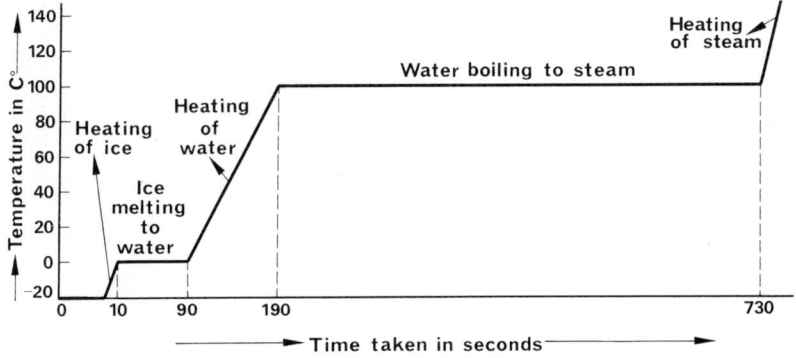

Figure 18.1. This graph plots the temperature against time as a steady amount of heat energy in joules per second is given to a certain fixed mass of ice causing it to change to water and later to steam.

127

is this heat energy which causes the *change of state* from ice to water—to bring about a more rapid molecular movement. Each kilogramme requires a certain quantity of heat energy measured in joules to change from ice to water and this is called the *specific latent heat of fusion of ice*. The word 'latent' means hidden, and is so used here because one cannot observe any rise in temperature as the heat is absorbed by the ice during the process of melting.

After all the solid ice has turned into liquid water the addition of more heat raises the temperature of the water and causes more rapid molecular movement again. It is the specific heat capacity of water which determines how rapidly the temperature of the water rises during the period from the melting point of ice to the boiling point of water. As we have seen from the table water has a relatively high specific heat capacity, higher than other liquids and solids, and this explains why water is suitable for use in the convectors of heating systems and in 'hot-water' bottles. It also explains why places near the sea are so popular in very hot weather. The earth, whose specific heat capacity is about one-fifth that of sea water, may become uncomfortably hot in the heat of the midday sun, but the water of the sea takes longer to heat and so remains cooler than the earth. The hot earth during the day heats the air immediately above inland places much more than the cooler sea is able to heat the air above the sea. As a result the cooler air from the sea blows in across the seashore and replaces the warmer lighter air that rises over places inland. During the night the breezes blow in the opposite direction because the land has less heat to lose and thus cools more rapidly than the sea. Places a long way inland experience much greater changes of temperature during the day and night than places near the sea because of this difference between the specific heat capacities of the earth and the sea. Such places inland have, for the same reason, what is known as a 'continental' climate— extremely hot summers and extremely cold winters.

When water is heated to its boiling point at 100° C another change of state takes place as more heat is given to it. Each kilogramme of boiling water needs much more 'hidden' heat to change it to its gaseous state as steam. You can

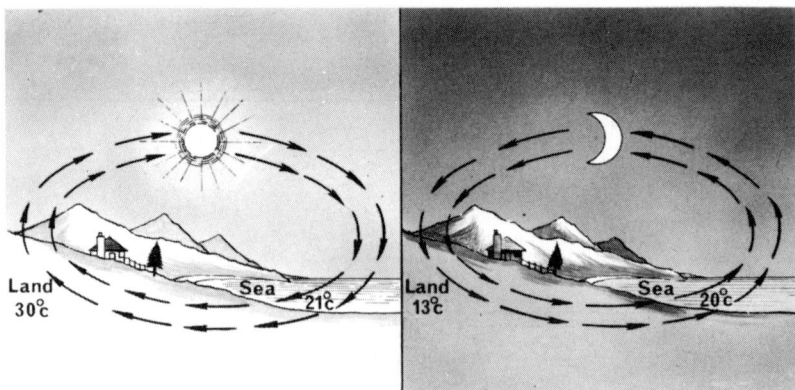

Figure 18.2. A sea breeze blows towards the land by day. A land breeze blows out to sea by night.

judge that is so by examining Figure 18.1 and noting the relatively long time it takes to convert the water to steam. This heat is known as the *specific latent heat of vaporization of water*. Compare the time taken to heat ice cold water until it starts to boil with the time taken by the water when it first starts to boil until it has all boiled away to steam.

Steam has a *specific heat capacity* and it can be heated above 100° C forming what is known as *superheated steam* found inside the boilers of steam engines.

At each stage in the process of change of state—ice to water to steam—the molecules gain more and more energy, move faster and faster, until, in steam, the paths traversed by the molecules have increased considerably. In reverse, when steam changes to water to ice, heat is given out and the molecules move slower along shorter paths.

Pure substances. Most pure substances have well defined melting points. In fact, scientists can often identify a substance by finding its melting point. In the same way they can discover if a certain substance is in a pure condition or whether it has been mixed with another substance. The addition of salt to water lowers its melting point. That is why snow on roads is sprinkled with salt when the temperature is just below the freezing point of pure water. The snow–salt mixture melts because the actual temperature is higher than the melting point of the mixture, and thus the roads become free from slippery snow surfaces. The addition of a chemical impurity called 'antifreeze' to the water in the cooling system of a motor-car prevents the water from freezing in cold weather. Motorists know that they can prevent the water freezing at much lower temperatures if they add more anti-freeze to the mixture in the system.

Mixtures. There are some mixtures like tar, pitch, glass, and wrought iron which, when heated, become soft and continue to melt over a range of several degrees rise in temperature. This is fortunate because, for example, it enables

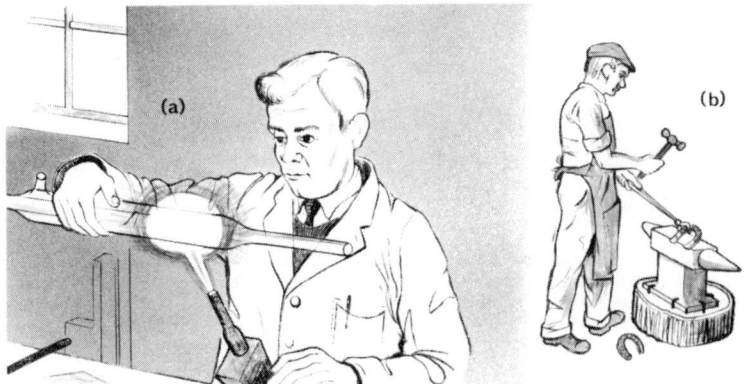

Figure 18.3. Skilled craftsmen (*a*) forming a glass condenser for a laboratory distillation plant, and (*b*) forging a horseshoe.

us to manipulate glass, to blow it, and to bend it whilst it is neither a true solid nor a true liquid. It also enables a blacksmith to shape wrought iron when it is in its semi-molten state and make iron gates and horseshoes.

Alloys. An alloy is a mixture of several metals. Solder is one of the most common of all the alloys. There are several grades of solder varying with the proportion of the various metals present in the alloy. Each grade has different physical properties such as hardness and melting point. Some alloys have extremely low melting points and these have special applications. The alloy called Wood's metal has a melting point of 65° C—it will even melt in a cup of hot tea! There are several alloys of bismuth with lead, tin, and cadmium that have very low melting points and they are sometimes used in automatic fire alarms and water sprinkler systems. The heat from an accidental fire in a building melts the metal alloy so that it releases a jet of water from a nozzle and at the same time rings an electric alarm.

The heat produced by an excessive current in an electric circuit may cause damage by fire. A fuse wire made of an alloy of tin and lead is included in the circuit. As this has a lower melting point than the copper wires in the rest of the circuit the fuse melts and breaks the circuit when the current exceeds a certain value. It does this before the copper wires get hot enough to cause a fire.

High melting points. Materials with high melting points have their uses also. Tungsten has the highest melting point of all the metals, namely 3 380° C, and is therefore used in electric lamps to make the filaments that must not melt when they become white hot.

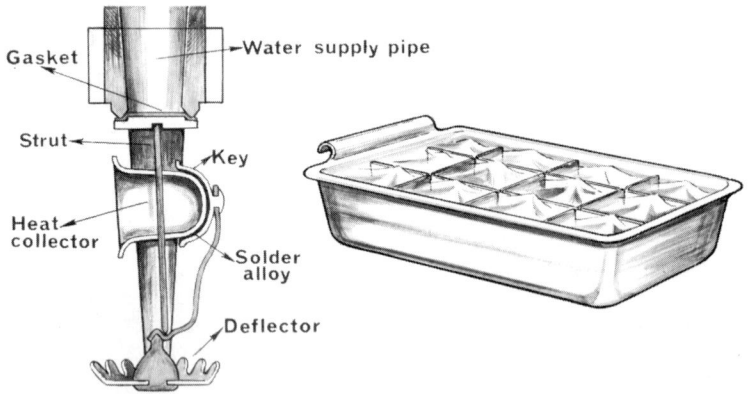

Figure 18.4. The cross-section of a water sprinkler. The soft metal gasket is held in position by the strut until the alloy between the heat collector and the key melts. When this happens the water pressure in the pipe above pushes everything aside and throws a jet of water against the deflector below. This forms a fine water spray which is spread around to put out the fire. Why are sprinklers put on ceilings and not on walls or floors?

Figure 18.5. This tray was filled to the brim with water and then placed in a refrigerator until the water in each section had turned to ice. The amount of expansion as the water solidified is shown by the humps formed on the free top surface of the water.

Fire doors, fire escapes, and safes are made of iron and steel because no ordinary fire can burn through iron (melting point of 1 500° C) or through steel (melting point of 1 400° C). It requires a very hot oxy-acetylene flame to melt these metals.

Change in volume. The change of state at the melting point is nearly always accompanied by a change in volume. Most substances expand on melting and contract on solidifying—gold, silver, and wax are examples. This is why gold and silver coins cannot be cast in moulds: they have to be stamped out. Beeswax also does not take the shape of a mould exactly and always forms a hollow in the middle of its surface as it sets.

Cast iron, type metal, and water contract on melting and expand on solidifying. Cast iron and type metal can be made into good castings. A hollow mould is prepared and the molten metal is poured in until it is full. As the metal solidifies it expands and takes the exact shape of the mould, filling every little space.

The expansion of water when it freezes can do a lot of damage; however in countries where the ground freezes in winter the force due to the expansion can help the farmer and gardener by smashing large clods of earth into small pieces. If there is no room for the water to expand it creates a *tremendous force*. This force bursts water pipes and motor-car engine blocks if water freezes in them during a frost; it breaks rocks to pieces in a process called weathering; it cracks concrete roads if the water can find its way under them.

If you examine a metal bucket of water, that has been left for some time in a temperature below the freezing point, you will find that the top surface of the ice formed has been pushed upwards, and the metal base of the bucket has been bent downwards by the large force of expansion created by the water on freezing.

Because of this expansion on freezing, ice is less dense than water and it

Figure 18.6. Icebergs have irregular shapes and jagged edges. They often rise several hundred feet above the sea. Why do they take a long time to melt?

therefore floats on water. Icebergs float and drift a long way before they melt. Only about one-tenth of the volume of an iceberg is above the surface of the ocean, the remaining nine-tenths are below and hidden from view. It is not surprising therefore that ships crossing the oceans where icebergs are to be found take extreme care and plot courses to avoid their tracks. An international icepatrol in the north Atlantic ocean regularly warns navigators of the positions of icebergs.

If ice were more dense than water it would sink as it formed. Lakes and ponds would then freeze solid when the temperature fell below the freezing point and remained there for a long time. As a result nearly all the animal life in these lakes and ponds would be destroyed. As it is, ice floats on water and after some days the surface of a lake or pond becomes a hard thick layer of ice.

The boiling point, like the melting point of most pure substances under normal conditions, is well defined and does not vary during the change of state from liquid to gas. If you are trying to identify a substance, it is often helpful to find its boiling point. An *impurity* in a liquid causes its boiling point to change. When salt is added to water it boils at a higher temperature. You can cook an egg more rapidly by boiling it in salt water than in ordinary water.

A *change of pressure* alters the boiling point of water—a decrease of pressure lowers the boiling point and an increase raises the boiling point.

Those who live on high mountains know well that potatoes and other vegetables cannot be cooked properly in boiling water contained in open vessels at those heights. Tea cannot be brewed or an egg boiled satisfactorily at very high altitudes. On the top of Mont Blanc, the highest mountain in Europe, water boils at 85° C, whereas at sea-level it boils at 100° C. A pressure cooker can, however, be used to cook anything successfully at any altitude; at low altitudes and at sea-level it will enable cooking times to be greatly reduced.

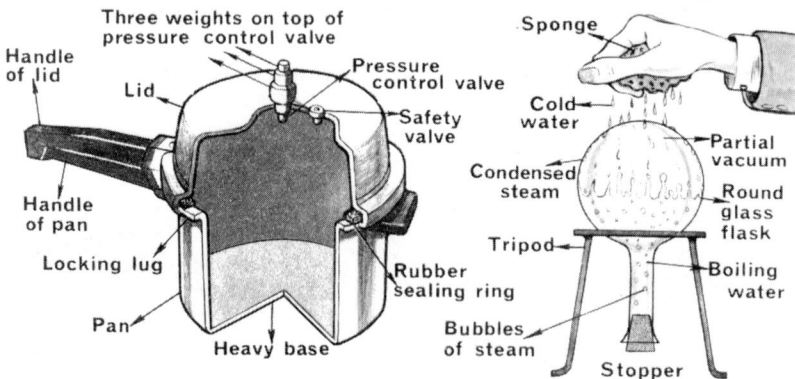

Figure 18.7. A food pressure cooker. Notice the strong body and lid. Why is a safety-valve fitted in the lid? How are different pressures controlled in the cooker when in use? What device seals the lid to the body of the cooker?

Figure 18.8. An experiment performed by the American scientist Benjamin Franklin. The water in the flask was boiled normally at 100° C, the steam drove out the air, the top was sealed with a cork, the flask inverted, cold water was poured on the outside to condense the steam inside the flask, and the water started to boil again at less than atmospheric pressure and at a temperature less than 100° C.

Pressure cookers have lids that can be fastened down securely to retain the steam escaping from the liquid. The resulting steam pressure causes the boiling point of the water inside to rise. Potatoes will cook at these pressures and temperatures in less than 10 minutes. The pressure in steam boilers used with engines and turbines is often so great that the water boils in them at 150° C.

The change of state from liquid to gas requires large quantities of heat to be supplied in the form of latent heat. This latent heat must be taken from the surroundings of the liquid. Change of state by boiling is a quick process when compared with the change of state by evaporation.

Boiling takes place throughout the whole of the liquid as well as on its surface but only when its temperature is at the boiling point of the liquid.

Evaporation takes place only on the surface of the liquid and occurs at any temperature. Evaporation takes heat from the surroundings of the liquid if no external source of heat is supplied.

The cooling effect of evaporation also explains why you feel cold as you stand about still wet after swimming, or if you are perspiring, especially if there is a wind blowing. This feeling disappears once you have removed the moisture from your body with a towel.

The rate of evaporation of a liquid varies according to several factors all of which you can observe when performing simple experiments:

1. *The area of the exposed surface.* Does a wet sheet dry more quickly when it is opened out on a line or when it is left folded? If it does, then the greater the area the greater the rate of evaporation.

2. *The temperature of the liquid.* Do wet clothes dry faster on a warm day than on a cold day, all other conditions being equal? If they do, the higher the temperature the greater the rate of evaporation.

3. *The rate at which the vapour is removed.* Do wet clothes dry faster on a

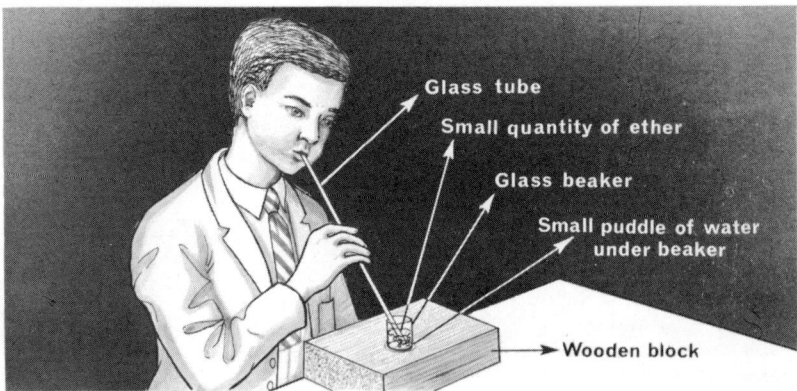

Figure 18.9. This boy is forcing the liquid in the beaker to evaporate quickly by blowing air through the fine nozzle at the end of the glass tube. He is giving a greater surface to the ether. The ether takes in latent heat from the glass beaker and the puddle of water below. Soon the beaker becomes covered with dew and the water freezes sealing the beaker to the wooden block.

windy cold day than on a calm cold day? If they do, the rate of evaporation is increased when the vapour is removed quickly.

4. *The nature of the liquid.* Put a few drops of petrol, water and oil on your hand. Do you notice different rates of evaporation? Which drops feel the coldest as they all evaporate?

The molecular theory of matter can provide an explanation for these observations. Consider the molecules near and forming the surface of a liquid. The molecules move about in all directions at varying speeds. From time to time one of the fastest molecules that happens to be travelling in the correct direction shoots up and out of the surface and then on and out of the region where it is attracted by those that remain behind. It escapes and becomes part of the vapour of the liquid. This is evaporation. The higher the temperature of the liquid the more fast ones there are to escape. Some of the less fast moving molecules that just do not get away can be made to escape by blowing across the surface of the liquid. If the number already escaped and present in the vapour above the liquid is great then a few of these return to the liquid. The vapour is said to be 'saturated' when the number of those escaping from the liquid every second equals those returning to it. The diagram in Figure 18.10 represents a liquid-vapour surface. There is no regular pattern. The molecules move in and out of the surface all the time and those shown are only a few of the millions present both in the liquid and in the vapour above.

The refrigerator uses the cooling effect of evaporation to keep the coils of the freezing unit cold. A liquid 'refrigerant' is cooled and pumped through a fine nozzle where it changes to its gaseous state. In doing so it takes in latent heat from the cooling coils of the freezing unit. The refrigerant, then in its gaseous state, is compressed back to a liquid giving up its latent heat to an outside

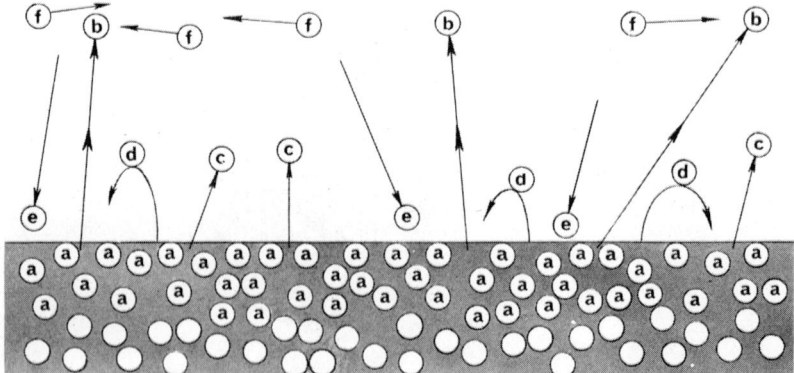

Figure 18.10. Evaporation. (*a*) Molecules moving about near the surface of the liquid. (*b*) These molecules escape completely because they are the fastest moving ones. (*c*) These molecules are not so fast as molecules (*b*) but they can be blown away by a strong air current. (*d*) These molecules are much slower than the others and are soon attracted back again. (*e*) These molecules are returning to the liquid from the vapour above. (*f*) These molecules are moving about and form the vapour, they travel long distances and only occasionally collide with another or get near enough to the surface to be attracted in again.

convector. The cycle of operations continues until the required temperature inside the whole refrigerator cabinet is reached and then the thermostat control switch turns off the supply of electricity.

When the reverse process takes place and a gas or vapour changes its state to a liquid the large amount of 'specific latent heat' it possesses is given out. Hence you have to be extremely careful not to allow any part of your body to come in contact with steam because you will be 'scalded' by the latent heat given out as the steam condenses to boiling water. Boiling water is hot enough and gives sufficient heat in cooling to hurt your skin, but steam gives off very much more heat when it condenses.

The process of condensation is the reverse of evaporation. In other words a vapour, formed by boiling or evaporationg a liquid, can be reconverted to its liquid state by taking away its latent heat. This is done by passing the vapour through a 'condenser' maintained at a temperature below the boiling point of

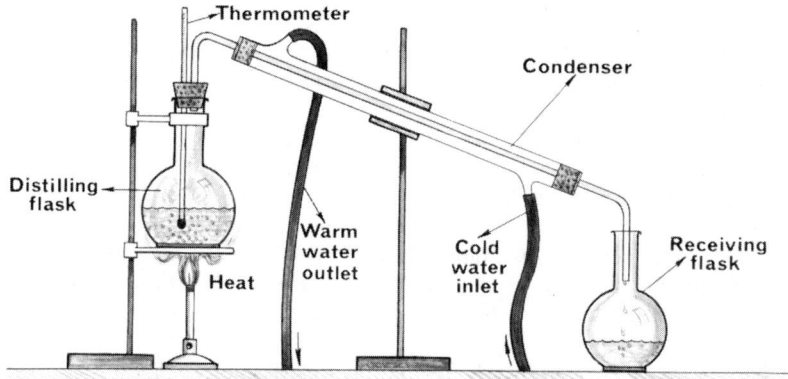

Figure 18.11. A distillation apparatus set up in a chemical laboratory to separate one substance from another. Industrial distillation plants operate on the same principles.

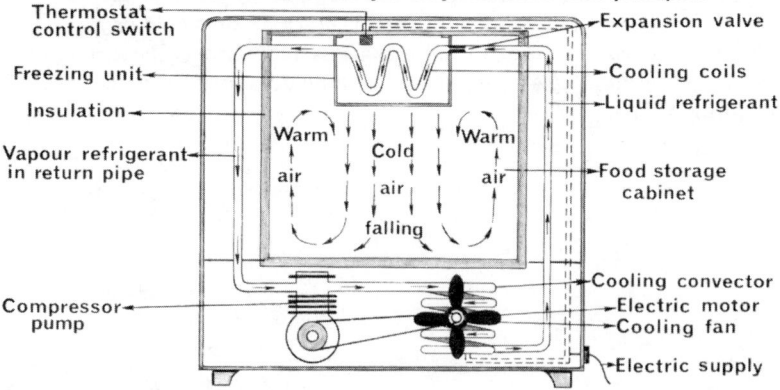

Figure 18.12. An electric refrigerator. The refrigerant is usually sulphur dioxide or 'Freon' and is pumped through the fine nozzle in the expansion valve where it turns to a gas in the cooling coils of the freezing unit. The refrigerant circulates in a sealed circuit and is continuously changing state. The heat absorbed in the food storage cabinet is thus liberated in the coils of the cooling convector outside and below the refrigerator.

K

the liquid. It has an important practical use as part of the process of *distillation*. Distillation is the process of first boiling a liquid and then condensing its vapour. It is used to separate the components of a mixture or to purify liquids. One component can be boiled off and its vapour condensed, while the others or the impurities are left behind. In this way petroleum is successively distilled to give petrol, paraffin, fuel oil, and many other products. Water is distilled to produce pure water for accumulators and chemical experiments.

Questions on chapter 18

1. What is happening when a liquid is boiling? What is meant by its boiling point? How is the boiling point changed when the pressure over its surface is increased?

2. Describe an experiment to show that if water is cooled sufficiently in an enclosed space it can exert a great force on freezing.

3. Explain the following: (a) water may be kept cool in a porous container which is placed in a draught, (b) steam at 100° C can cause much worse scalding than water at 100° C, and (c) water pipes sometimes crack in very cold weather.

4. What happens to the molecules when: (a) ice melts, (b) water warms up, (c) water evaporates?

5. It is believed that all substances consist of molecules. Say how the molecules behave during the evaporation of a liquid, describing what happens when the space above the liquid contains (a) unsaturated vapour, and (b) saturated vapour.

6. A boy noticed that on a hot, sultry day he felt much cooler when he sat before a spinning electric fan. He measured the temperature of the air coming from the fan and found it to be no cooler than the air elsewhere in the room. Explain why he felt cooler.

7. Type metal is an alloy which, when it solidifies, forms a very clear impression of the mould in which it is placed. Explain why it does this and state whether water or paraffin wax will do the same.

8. Why is it sometimes difficult to form snowballs on a cold day?

9. Water lubricates the blade of the skate as it moves over the surface of the ice during skating. Explain why the same thing does not take place under the ski during skiing.

19. The measurement of heat

Heat is a form of energy. In the last chapter we learned that if we give a body heat energy we increase the energy of its molecules by increasing their speed of movement. We learned also that scientists base their measurements of heat energy on the change in temperature this increase of energy produces in a certain quantity of water. Temperature, as we have learnt, is the property of a body which determines whether heat will flow to or from a particular body with which it is placed in contact.

The *joule* is the unit of heat, energy, and work. Two other units of heat are in common use because this is the way measurements in heat have developed. They are related by the equations:

$$4.2 \text{ joules} = 1 \text{ calorie}$$
$$4\,200 \text{ joules} = 1 \text{ Calorie}$$

$$\left.\begin{array}{l} 1 \text{ calorie} \\ 1 \text{ Calorie} \end{array}\right\} \begin{array}{c} \text{is the amount of heat} \\ \text{necessary to raise the} \\ \text{temperature of} \end{array} \left\{\begin{array}{l} 1 \text{ g of water } 1° \text{ C.} \\ 1 \text{ kg of water } 1° \text{ C.} \end{array}\right.$$

The *joule* is the unit used by scientists in their laboratories and is accepted as an international unit. The *Calorie*, the kilogramme calorie, or the large calorie is 1 000 times as large as the calorie and is used in the measurement of the heat value of foods. It is also used internationally.

A *calorimeter* is an instrument used to measure the quantity of heat available when a fuel is burnt or a foodstuff consumed. The quantity of heat available in a certain mass of such a substance is known as its *calorific value*.

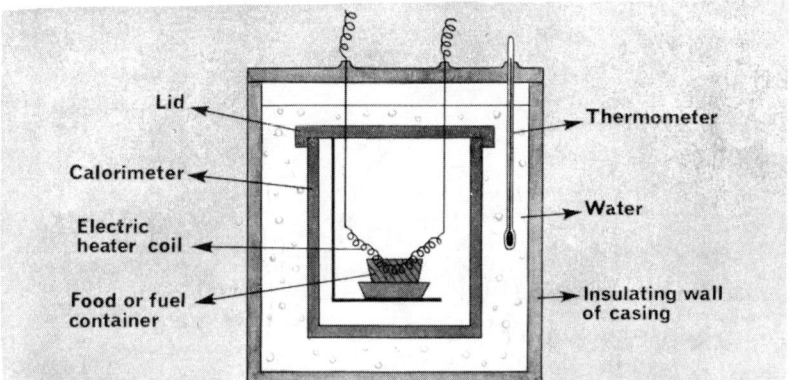

Figure 19.1. A fuel calorimeter. The quantity of heat emitted by the burning food or fuel is determined by the rise in temperature of the water surrounding the sealed calorimeter. The burning is started by the electric heater coil immersed in the fuel container.

137

The calorific values of some fuels and foodstuffs are:

Coal	..	..	..	..	8 Calories per gramme
Petrol	..	..	..	..	11 Calories per gramme
Bread	..	..	..	..	2.6 Calories per gramme
Beef	..	..	..	..	2.0–3.0 Calories per gramme
Fats	..	..	..	;.	7.5–9.0 Calories per gramme

We can compare the heat available in different metals at the same temperature when they are allowed to cool to air temperature by performing the following experiment and observing the results. Six cylinders of brass, iron, copper, lead, tin, and zinc are made each of the same mass and diameter and each fitted with a small hook at one end. They are heated to the same temperature in boiling oil, quickly removed and wiped one by one, and placed on one of the grooves of a sloping wooden board (see Figure 19.2). Mounted a small distance off the board is a sheet of wax foundation used by beekeepers to encourage bees to build regular cells in their hives. The cylinders melt their way as they descend into the wax foundation. Those cylinders having the greatest heat available when cooling from the temperature of the boiling oil to the air temperature sink the deepest into the wax foundation. This experiment gives an approximate value of the relative capacity that the different substances possess of giving out heat, or in other words their *specific heat capacities*. It can only be approximate because some heat is lost to the air and to the wooden board, and all metals do not conduct heat at the same rate from the body of the metal to the advancing surface at the bottom of the cylinder that is melting its way through the wax.

The result of this experiment can be expressed by saying that the depth of the wax melted depends on the quantity of heat given out by the cylinder, that is, on the specific heat capacity of the metal.

Another experiment can be used to tell us more about the heat needed to raise the temperature of a metal body, and that means also the heat available

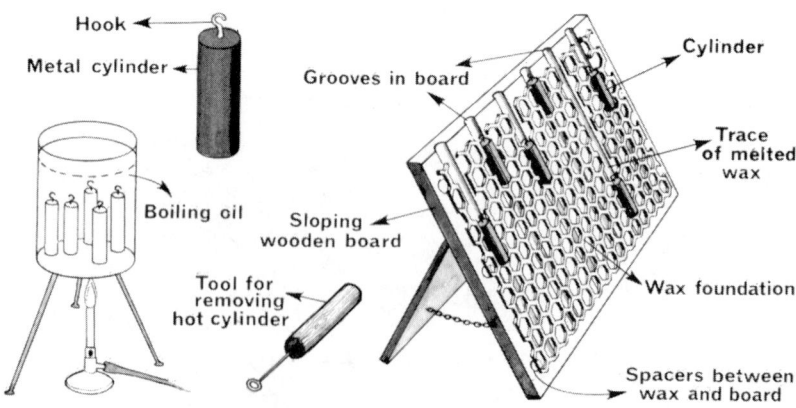

Figure 19.2. An experiment to compare the specific heats of different metals.

to be given out as it cools. Thick metal cans have to be prepared all having the same mass and all covered with a heat insulator such as felt or polystyrene. Into each can is poured the same quantity of boiling water. After a short time the cans have absorbed the heat from the water and the final temperature of the cans and the water becomes stationary. Those metals absorbing the greatest quantity of heat from the water, in other words those having the highest specific heat capacity, lower the temperature of the water the furthest.

Liquids, too, have different specific heat capacities. This can be shown by pouring equal masses of water and another liquid, heated to the same temperature, into two similar thick metal cans and conserving the fall in temperature of the water and liquid under these same conditions. Water has the highest specific heat capacity of all ordinary substances.

In the process of performing these last two experiments one of the substances loses heat and another one gains heat. Provided that no heat escapes outside to other bodies then the heat lost by one must equal the heat gained by the other. This process is known as the *method of mixtures* because substances are mixed or placed together so that the heat is redistributed between them. The following equation always applies:

$$\begin{array}{cc} \text{HEAT LOST} & = & \text{HEAT GAINED} \\ \text{by the hot bodies} & & \text{by the cold bodies} \end{array}$$

It is convenient to express all these relationships in the form of an equation thus:

$$\begin{array}{ccccc} \text{QUANTITY} & = & \text{MASS} & \times & \text{TEMPERATURE} & \times & \text{SPECIFIC HEAT} \\ \text{OF HEAT} & & \text{OF BODY} & & \text{CHANGE} & & \text{CAPACITY} \end{array}$$

(in joules) (in kilogrammes) (in degrees C)   (in J kg^{-1} °C^{-1})

or Q $=$ m $\times$ $(T_1 - T_2)$ $\times$ c

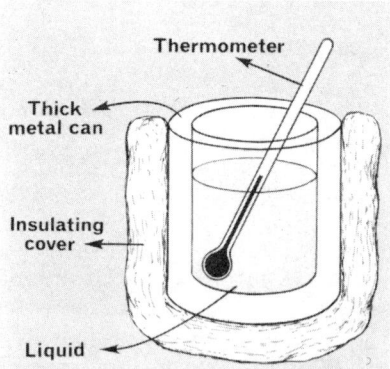

Figure 19.3. The 'method of mixtures' in operation. A hot liquid sharing some of its heat with a cold thick metal can.

Example 1. How many joules are needed to raise 1 kilogramme of water from 25° C to its boiling point? How long will this take if an electric immersion heater is used that supplies 315 joules per second?

$$Q = m \times (T_1 - T_2) \times c,$$

$$\therefore \; Q = 1 \times (100 - 25) \times 4\,200, \quad \text{(see page 127)}$$

$\therefore$ Quantity of heat needed $= 3.15 \times 10^5$ joules

$$\text{Time taken} = \frac{\text{Quantity of heat}}{\text{Rate of supply of heat}},$$

$$\therefore \; \text{Time taken} = \frac{3.15 \times 10^5}{315},$$

$\therefore$ Time taken $= 1000$ seconds.

The latent heat required by a substance to change its state from a solid to a liquid is known as the *specific latent heat of fusion*. It is measured in joules per kilogramme and is the quantity of heat required to change the state of one kilogramme of the substance without change of temperature. For the change from ice to water this quantity is 3.3×10^5 J kg^{-1} and is a fairly high value and that is why the ice on a pond takes a long time to form, sometimes many days, and a long time to disperse.

When the change is from a liquid to a gas the *specific latent heat of vaporization* is involved. In the same way this is measured in joules per kilogramme. For the change from water to steam this is 2.2×10^6 J kg^{-1}, and therefore to boil a kettle of water dry takes a relatively long time. This also explains why in some electric power stations enormous cooling towers are built, standing hundreds of feet high, to condense the steam withdrawn from the steam turbines that spin the generators. So much heat has to be taken from the steam when it condenses that large sprays of cold water are needed. In some countries this heat is not wasted for it is used to supply hot water to houses, factories, and other buildings.

The two specific latent heat relationships can be expressed by simple equations thus:

QUANTITY OF HEAT	=	MASS OF SUBSTANCE CHANGED STATE	×	SPECIFIC LATENT HEAT OF FUSION OR VAPORIZATION
(in joules)		(in kilogrammes)		(in J kg^{-1})
or Q	=	m	×	l

It is usual to use the same symbol l for the latent heat whether it refers to the specific latent heat of fusion or of vaporization because we are unlikely to confuse their meanings or values in our calculations.

Example 2. A large iced drink is prepared for a party by adding a mass of 2 kg of ice at 0° C to 16 kg of a diluted fruit juice at 20° C in an insulated container of negligible heat capacity. If the specific heat capacity of the diluted fruit juice is $4\,000\ \text{J kg}^{-1}\,^{\circ}\text{C}^{-1}$ what is the final temperature of the iced drink?

The ice gains heat and the diluted fruit juice loses the same amount of heat because no heat is lost from the mixture.

$$\text{Heat gained by ice in melting} = m \times l$$
$$= 2 \times 3.3 \times 10^5 \qquad \text{(see page 140)}$$
$$\text{Heat gained by ice after melting} = m \times (T_1 - T_2) \times c$$
$$= 2 \times (T_1 - 0) \times 4\,200 \text{ (see page 127)}$$
$$\text{Heat lost by diluted fruit juice} = m \times (T_3 - T_1) \times c$$
$$= 16 \times (20 - T_1) \times 4\,000$$

Equating these two changes in heat:
$$(2 \times 3.3 \times 10^5) + (2 \times T_1 \times 4\,200)$$
$$= (16 \times 20 \times 4\,000) - (16 \times T_1 \times 4\,000)$$
or $\qquad (8\,400 + 64\,000)\,T_1 = 128 \times 10^4 - 66 \times 10^4$
$$T_1 = 8.6^{\circ}\ \text{C}.$$

Therefore the final temperature of the iced drink when all the ice is melted is 8.7° C.

The *heat capacity* of a container is the heat required to raise its temperature by 1° C and is measured in $\text{J }^{\circ}\text{C}^{-1}$.

Example 3. A kettle containing 1 kg of boiling water at 100° C is emptied on to an ice-covered pavement at 0° C. How much ice is melted if all the heat of the boiling water melts ice so that it all finishes as water at 0° C and no heat is lost to the air or the pavement itself?

The boiling water cools from 100° C to 0° C.

If m g of ice melts without a change in temperature.

$$\text{Heat gained by ice} = m \times l,$$
$$= m \times 3.3 \times 10^5 \text{ J.}$$
$$\text{Heat lost by water} = m \times (T_1 - T_2) \times c,$$
$$= 1 \times (100 - 0) \times 4\,200,$$
$$= 4.2 \times 10^5 \text{ J.}$$

But heat gained by ice $=$ Heat lost by water,

or $\qquad m \times 3.3 \times 10^5 = 4.2 \times 10^5,$
$$\therefore\ m = 1.27 \text{ kg.}$$

$\therefore$ The boiling water melts 1.27 kg of ice on the pavement.

Sometimes the calculation we have to make is more complicated than the one shown above in Example 3. This is because the bodies that gain or lose heat are in containers and the containers also gain or lose heat so that we then have to add together the two quantities of heat gained or lost.

Example 4. 500 g of water at 20° C contained in an aluminium tray of mass 200 g and specific heat capacity $900\ \text{J kg}^{-1}\,^{\circ}\text{C}^{-1}$ are placed in the cooling unit of a refrigerator. If the cooling unit can extract $42\ \text{J s}^{-1}$ from the ice tray and water, how long will it take to freeze all the water in the ice tray?

Heat lost by water in cooling from 20° C to its freezing point at 0° C.

$$= m \times (T_1 - T_2) \times c,$$
$$= 0.5 \times (20 - 0) \times 4\,200,$$
$$= 4.2 \times 10^4 \text{ J}.$$

Heat lost by ice tray in cooling from 20° C to the freezing point of water at 0° C

$$= m \times (T_1 - T_2) \times c,$$
$$= 0.2 \times (20 - 0) \times 900,$$
$$= 3.6 \times 10^3 \text{ J}.$$

Heat lost by the water at 0° C in changing to ice at 0° C

$$= m \times l,$$
$$= 0.5 \times 3.3 \times 10^5,$$
$$= 1.65 \times 10^5 \text{ J}.$$

∴ Total heat lost by the water and ice tray

$$= 4.2 \times 10^4 + 3.6 \times 10^3 + 1.65 \times 10^5$$
$$= 2.1 \times 10^5 \text{ J}$$

∴ Time taken to freeze the ice $= \dfrac{2.1 \times 10^5}{42}$ s

$$= 83.3 \text{ min.}$$

∴ It takes 83 minutes to cool the water in the ice tray and then to turn the water into ice.

Example 5. How long will it take to boil a kettle of water dry if it takes 4 minutes to bring the water in the kettle to boiling point from 20° C?

Let us assume that the kettle has a small mass and that it contains a large mass of water so that the heat needed to raise the temperature of the kettle itself can be neglected in this calculation.

Heat needed to raise the temperature of the water from 20° C to 100° C if the kettle contains m kg of water

$$= m \times (T_1 - T_2) \times c$$
$$= m \times (100 - 20) \times 4\,200$$
$$= m \times 3.36 \times 10^5 \text{ J}.$$

Figure 19.4. One of the cooling towers of a British thermal power station. These are often built near rivers in order that plenty of water is available.

Heat needed to change the m kg of boiling water at 100° C in the kettle to steam at 100° C

$$= m \times l,$$
$$= m \times 2.2 \times 10^6 \text{ J}.$$

To supply $m \times 3.36 \times 10^5$ J to the water took 4 minutes,

∴ To supply $m \times 2.2 \times 10^6$ J to the water will take

$\dfrac{4 \times m \times 2.2 \times 10^6}{m \times 3.36 \times 10^5}$ minutes or about 27 minutes.

∴ To boil the kettle dry will take (4 + 27) or 31 minutes.

Questions on chapter 19

1. Which would make the better and more effective bed warmer, a hot brick of mass 2 kg or a rubber bag containing 2 kg of hot water if they were both at the same temperature? Explain the reasons for your choice.

2. Which of the following materials do you think would give out the greatest quantity of heat in cooling through 80° C, (a) 500 g of iron, (b) 500 g of lead, (c) 500 g of copper, (d) 500 g of rock, or (e) 500 g of water? What do you require to know before you can be sure that the answer you give is the correct one?

3. Strawberries, raspberries, tomatoes, oranges, fruit juices, and milk all have higher specific heat capacities than foods like butter, cheese, fats, and meats. Which group are more costly to refrigerate when stored in large quantities? Explain why the foods in one group will take longer to warm up when taken out of the refrigerator than the other.

4. What is the heat capacity of a metal container at 20° C if when 1.2 kg of boiling water is poured into it the final temperature of the water is observed to be 80° C. The specific heat capacity of water is 4 200 J kg^{-1} °C^{-1}.

5. During the process known as the method of mixtures a metal of mass 0.5 kg and specific heat capacity 126 J kg^{-1} °C^{-1} is heated to 100° C. It is then put into 1.0 kg of a liquid in a container at 20° C and the final temperature is observed to be 27° C. The heat capacity of the container is 517 J °C^{-1}. What is the specific heat capacity of the liquid in J kg^{-1} °C^{-1}?

6. A large aluminium washer at 100° C and of mass 30 g is dropped on the surface of a block of ice at 0° C. Calculate the specific heat capacity of aluminium if 7.5 g of ice are melted. The specific latent heat of fusion of ice is 3.3×10^5 J kg^{-1}.

7. Buckets of water placed in a small cellar can help to prevent apples and tinned foods stored there from freezing in cold weather. Why must one change the water frequently?

8. What is the quickest way of evaporating a liquid if it cannot be boiled?

9. In a laboratory experiment a steady gas flame is placed under 100 g of water at 12° C and at the end of 220 s the water commences to boil. It takes a further 1 300 s to boil all the water away. Calculate a value for the specific latent heat of vaporization of water if the atmospheric pressure was normal.

10. 10 kg of water at 0° C are frozen into ice at 0° C. The specific latent heat of fusion of ice is 3.3 × 10⁵ J kg⁻¹.

State whether energy is released into or absorbed from the surroundings in the above change. Calculate the quantity of heat energy concerned.

11. Some crushed ice at a temperature below the freezing point of water was placed in a saucepan. The bulb of a thermometer was placed in the ice. The ice was then placed over a steady source of heat and the reading of the thermometer taken at regular intervals of time, until all the ice had melted and all the water formed was boiled away into steam.

Draw a graph showing the relation between the reading of the thermometer (temperature) and the time, throughout the experiment. Describe what happened in the various well-defined sections of the graph.

12. 10 g of dry steam at 100° C are conducted into a cavity in a large block of ice at 0° C. How many grammes of ice will be melted? (Specific latent heat of steam is 2.2 × 10⁶ J kg⁻¹, specific latent heat of ice is 3.3 × 10⁵ J kg⁻¹.)

20. Heat engines

Heat energy is converted into mechanical energy by means of a heat engine. There are many types of heat engine and there are many sources of heat energy available for use in these different engines, e.g. the steam reciprocating engine, the steam turbine, the petrol explosion engine, the Diesel fuel oil engine, and the jet combustion engine. The general principle of most of these heat engines is that a blade or a piston is acted upon by a force developed as a result of a pressure on its surface area.

In the piston engines the force on the piston pushes a rod and this sets machinery in motion. The movement of the rod takes place backwards and forwards in a straight line and this movement is then converted into a rotary movement by means of connecting rods, bearings, and cranks. Obviously for the engines with rotating blades no conversion to rotary movement is necessary.

The reciprocating steam engine has few moving parts and is a fairly cheap and simple engine. Steam under pressure is led to the steam chest of the engine where it enters the cylinder through one of the ports and exerts a pressure on one of the sides of the piston. The piston moves, operates a crank which turns the crankshaft and produces the rotation of the driving wheel and flywheel. An eccentric fitted to the crankshaft alters the position of the slide valve so that the steam can be cut off from this side of the piston and directed to press upon the other side at the correct instant, to force it back again. The vibrations created by the to-and-fro movement of the piston and its various connecting rods make this engine suitable for low speed work only. It has had much success in the past in driving small mechanisms such as shovels, cranes, and locomotives.

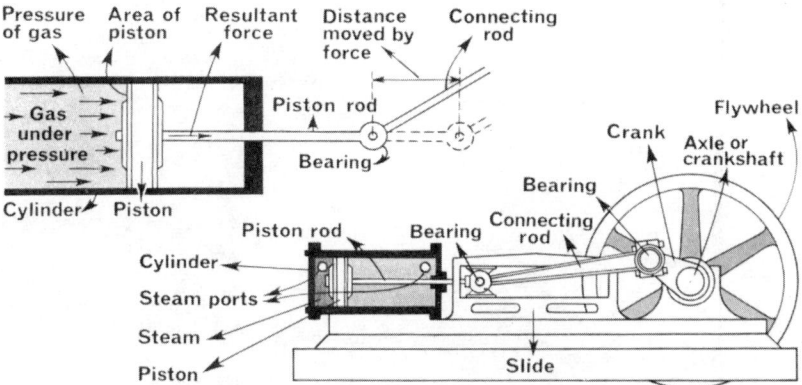

Figure 20.1. Work is done when gas under pressure pushes a piston a distance along a cylinder. The work done can be calculated for piston engines if the average pressure, the area of the piston, and the distance (length of stroke) are known.

Figure 20.2. This diagram shows how the movement of a piston backwards and forwards is able to cause rotary movement.

The steam turbine is more efficient than the steam engine, especially at high speeds when it is capable of converting more of the energy of the fuel into useful mechanical energy. The turbine is a rotary engine; high speeds are possible because it has no parts which move to-and-fro and thus vibrations are eliminated. Modern turbines consist of many series of blades attached to a rotating drum, called the rotor; between these blades are placed series of stationary blades, called the stator, attached to the casing. The surfaces of the two sets of blades are concave and they are placed opposite to one another. High pressure steam passes between the blades and turns the rotor. The speed of rotation is controlled by inlet valves leading to the steam jets.

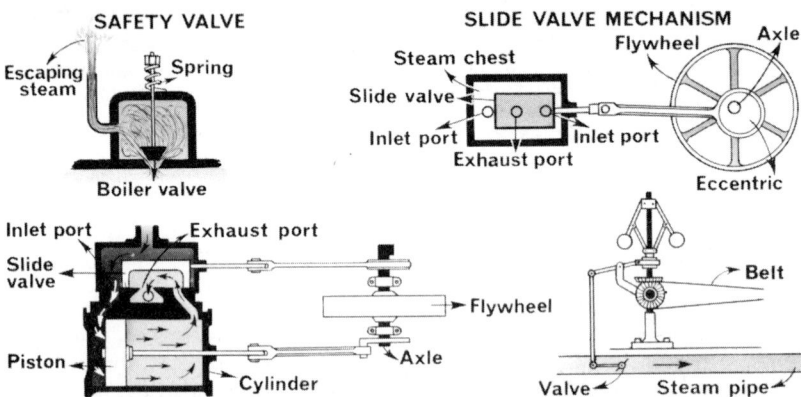

Figure 20.3. Sectional diagrams of a one cylinder reciprocating steam engine. The slide valve is moved by an eccentric attached to the main axle and flywheel. The safety valve, often fitted close to the boiler, opens before the steam pressure becomes too great. The governor regulates the speed of the engine by opening and closing the valve in the steam inlet pipe to the steam chest. The heavy flywheel maintains a steady speed of rotation by smoothing out the sudden pulses given at each stroke. The flywheel has enough energy stored in it because of its rotation to carry the engine over the 'dead points' at each end of the stroke of the piston.

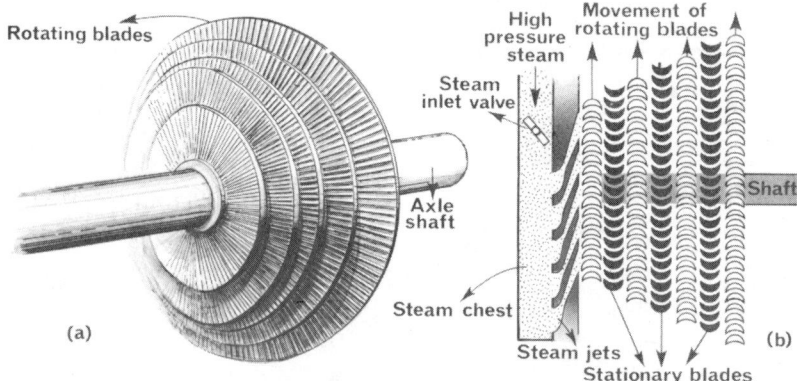

Figure 20.4. (a) The blades of the rotor of a steam turbine. (b) The method of mounting the rotating and stationary blades in a steam turbine. The steam expands as it passes from one pair of blades to the next so the diameters of the rotor, the stator, and the casing are made increasingly larger towards the steam outlet.

The steam needed for these engines can be supplied either from a *steam boiler*, or from a *nuclear reactor* and *heat exchanger*.

Fuel is burnt in the fire box or combustion chamber of a steam boiler. The hot gases and flames pass through a number of tubes fitted to increase the surface area which is in direct contact with the water to be converted into steam. The actual boiler surrounds the tubes and most of the fire box, and contains the water and steam. The fuel can be solid like coal or coke, liquid like heavy oil, or a gas.

Fissile material like uranium 235, splits up and releases energy in a nuclear reactor. The energy released is in the form of heat which changes a circulating cold gas or liquid to a hot one; this, in its turn, circulates through a heat exchanger where water is converted into steam, and the now cooled gas or liquid returns to be reheated.

When water changes to steam it increases its volume by about 1 700 times and the steam within the boiler or the exchanger often develops a high pressure of many atmospheres.

The internal combustion engine burns its fuel inside the engine itself and that is how it gets its name. Its efficiency varies considerably and in some cases it may be as high as 30%. The I.C. engine is generally lighter and more transportable than the steam engine and it is not attached to a boiler as the steam engine is. There are three general types of internal combustion engines.

1. The *petrol engine* derives its energy from an exploded mixture of air and the vapour of highly volatile petrol. It is often called the four-stroke cycle engine because the operations taking place during each of four strokes inside the cylinder repeat themselves. The strokes follow one another in this order: intake, compression, explosion, and exhaust. The mixture of air and petrol vapour which is drawn in is compressed to about one-sixth of its volume and is then exploded by an electric spark.

A *two-stroke cycle* engine by means of cleverly designed openings for the intake and the exhaust of gases combines the intake and compression strokes in one single stroke and the explosion and exhaust strokes in another. The

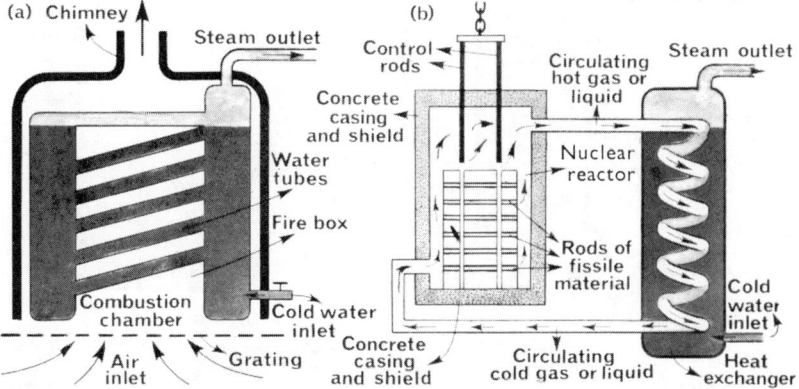

Figure 20.5. Methods of steam production. (*a*) The fuel steam boiler. (*b*) The nuclear reactor and heat exchanger.

two-stroke single cylinder engine runs more evenly than the four-stroke single cylinder engine because it gives a power stroke for every one revolution of the crankshaft whereas the four-stroke single cylinder engine gives a power stroke only once every two revolutions of the crankshaft. The shape of the top surface of the piston is so designed that it deflects the intake gases in such a way that they replace completely the exhaust gases in one movement.

2. The *Diesel engine* consumes a heavy fuel oil that is injected into a cylinder full of highly compressed and very hot air. The temperature of the air that has been compressed to about one-sixteenth of its volume is so high that the mixture explodes without the aid of an electric spark. In all other respects the cycle of operations follows the same pattern as in the four-stroke petrol engine. A two-stroke cycle Diesel engine that has proved highly efficient has been constructed in which the air blower provides intake air already considerably compressed before it enters the cylinder.

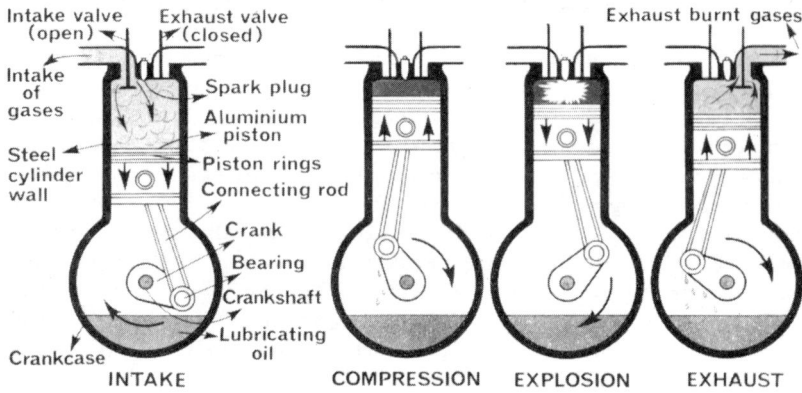

Figure 20.6. The four strokes of the petrol engine. This shows the positions of the valves controlling the openings to the cylinder, the direction of the piston movements, and the condition of the gases during each stroke.

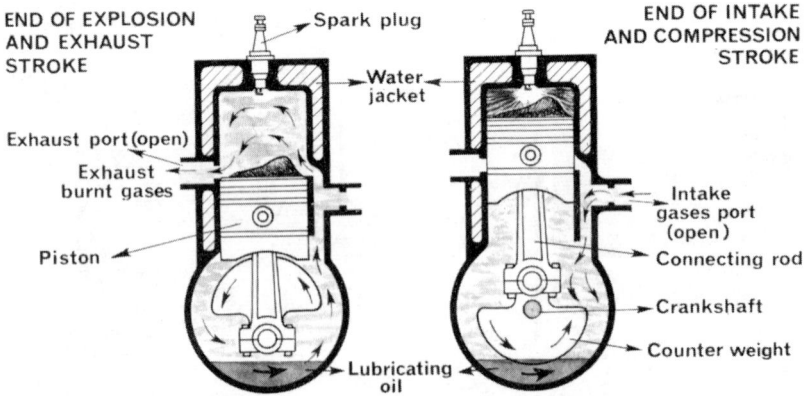

Figure 20.7. A two-stroke cycle engine. There is an explosion stroke every time the piston travels away from the cylinder head.

3. The *jet engine* ejects the products of combustion of its fuel backwards at very high speeds in order to produce a forward thrust. It is not necessary for the ejected small particles to push against anything as they are ejected; in fact, they leave the engine more easily and more rapidly if there is no air present and produce in this way a greater thrust. The thrust they produce is really a reaction to the product of their mass and their speed. Similarly if you are in a small boat and free to move on the surface of a pond and you start to throw stones out backwards, you and your boat will begin to move forwards. This operation is the same in principle—the faster you throw the stones the greater the speed the boat will attain. This explains why aircraft fitted with jet engines fly very high where the air is very thin and their efficiency is greatest.

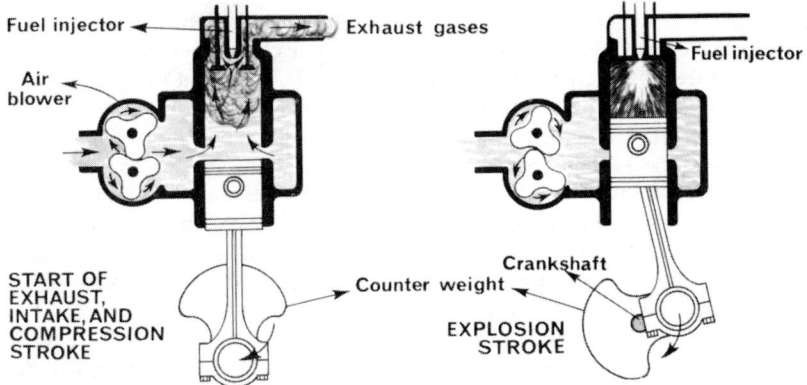

Figure 20.8. A two-stroke Diesel engine showing the exhaust, intake, and compression stroke about to start, and the explosion stroke. The diagram shows the fuel injector between the twin exhaust valves. The piston is such that it nearly reaches the top of the cylinder producing highly compressed gas there.

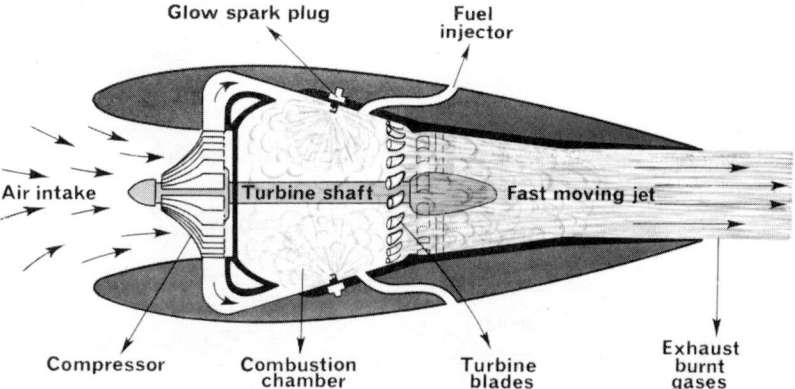

Figure 20.9. The turbo-jet engine. Air enters, is compressed, and driven into the combustion chamber where it meets the fuel and the two burn rapidly. The escaping gases rotate the turbine and then form the high-speed jet that develops the reaction or thrust on the engine. The turbine turns the compressor fitted on the same shaft. If a propellor is fitted to the shaft through gears then two kinds of thrust become available.

4. *Modern rockets*, which carry space ships beyond the attraction of the earth and which put artificial satellites into orbit around the earth, are propelled by jets using the principle of reaction. These rockets carry their own fuel supply and enough oxygen for its complete combustion. Sometimes they are built in sections so that as soon as the fuels are exhausted in one section it is released and the next section fires its jet.

We have spoken of the *efficiency* of the various heat engines. What does this mean? If all the heat released from the burning fuel were transformed into mechanical energy then the engine would be 100% efficient. But all engines lose a lot of their heat in one way or another—steam engines lose it in the heat present in the exhaust steam—steam turbines in the cooling towers—internal combustion engines in the exhaust gases and in the water circulating through the radiator—jet engines in the hot gases of the jet itself. Thus only a fraction of the heat energy available is transformed into mechanical energy.

What mechanical energy is equivalent to a certain amount of heat energy? This is a problem that has concerned many people in many lands; Count Rumford in Germany, Sir Humphrey Davy in England, the German physician Mayer, and the French chemist Regnault. Then finally the English physicist James Prescott Joule found by a series of experiments that a definite quantitative relationship existed between heat and mechanical energy. Joule used several methods in his experiments from all of which he found the amount of heat generated as a result of mechanical work being done. The relationship is known as the *mechanical equivalent of heat*.

We have already considered in Chapter 13 the mechanical unit of energy and work, namely the joule, and in Chapter 19 the unit of heat energy the calorie. It is the intention of scientists to replace the calorie by the joule as the basic unit of heat energy. The mechanical equivalent of heat relationship establishes the fact that 1 joule equals 0.24 calorie. 1 joule, as we remember, is 1 newton metre or the energy involved when a force of 1 newton moves through a distance of 1 metre.

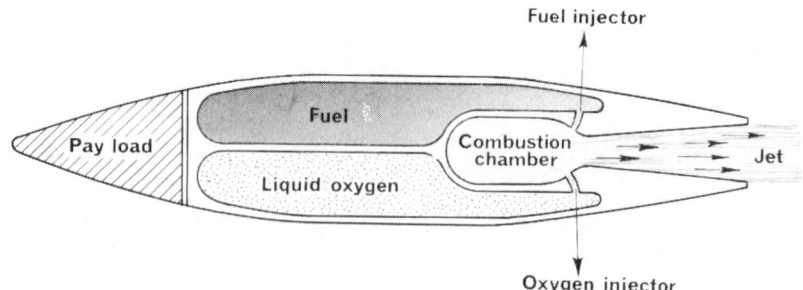

Figure 20.10. A modern rocket. The pay load weighs only a very small fraction of the total weight of the entire rocket. The diagram shows a single stage rocket only.

Later in this book, in Chapter 29, we shall consider the unit of electrical energy. It is also a joule and is the energy involved when a current of 1 ampere passes across a difference of potential of 1 volt for 1 second.

Thus mechanical, heat, and electrical energy are measured by the same unit, the *joule*.

Joule conducted many scientific experiments in the fields of mechanical, electrical, and heat energy, working alone in a private laboratory between the years 1843 and 1878. He was a Fellow of the Royal Society, a distinction that he richly deserved.

Once this mechanical equivalent of heat relationship had been definitely established it became clear that we cannot get out of any machine more energy than we put into the machine. Thus there can be no such thing as the perpetual-motion machine which was the goal of many fanciful inventors and designers in the past. They hoped to build a machine that, once started, would continue to run doing large quantities of work for the expenditure of little input energy.

The experiments performed by Joule made people look for other energy relationships and led directly to the scientific idea of the *Conservation of Energy*. This states that energy cannot be created or destroyed—it can only be transformed from one form to another. This is an important principle and until the work of Einstein and scientists in recent years this was accepted without question. Now it appears that we must agree that matter is also another form of energy because energy can be obtained by breaking up the atom and its nucleus. The atomic bomb is an example of the sudden generation of energy from the atom, and the nuclear power station shows how the energy of the atom can be controlled, liberated slowly, and used in the production of electrical energy for peaceful purposes.

L

Questions on chapter 20

1. Name all the sources of energy found on the earth that are derived from the energy of the sun. How are they stored? Are there any sources of energy that cannot be traced to the energy of the sun?

2. What advantages has a steam turbine over a reciprocating steam engine?

3. Why are condensers fitted to condense the exhaust steam from steam engines and turbines?

4. What is the function of an eccentric wheel in a steam engine?

5. Describe and compare in the working of a steam engine and a steam turbine: (a) speed of revolution, (b) efficiency, (c) reversibility.

6. List the advantages and disadvantages of the steam engine and the internal combustion engine.

7. Explain why the two-stroke cycle engine piston is irregularly shaped at the top.

8. Why must the crankcase fitted to a two-stroke cycle engine be gas tight?

9. Explain the essential reasons why a two-stroke cycle engine is simpler in construction than a four-stroke cycle engine.

10. Explain, with the aid of simple diagrams, the ways in which diesel engines differ from petrol engines in (a) the injection of fuel, and (b) the method of firing.

11. Why is the crankshaft of an engine turned more smoothly if driven by a 6 cylinder engine than by a similarly powered 4 cylinder engine?

12. What is a counterbalance weight? Describe where it is placed in an engine. Illustrate your answer with diagrams.

13. Explain why an internal combustion engine that has been excessively heated sometimes continues to run after the ignition has been switched off.

21. Waves

One can easily see the movements of some kinds of waves whereas it is impossible to see the movements of other waves. Waves travelling outwards from a disturbance over the surface of a pond or along a rope lying on the ground can be followed by the eye and their shape observed.

A wave travels away from the disturbance that created it and the *amplitude* or the displacement of the particles forming the wave itself gradually becomes smaller and smaller the further the wave travels. In this case if we look carefully at the particles that are disturbed by the passage of the wave we shall see that the particles are disturbed at right-angles to the direction that the wave is travelling. This is known as a *transverse* wave.

Water waves. Waves on the *surface of water* are usually made up of waves of many wavelengths due to a variety of causes—raindrops, winds, and varying depth of water. These waves travel at different speeds. When sea waves approach shallower water near the shore their speed depends only on the depth of water and thus they are all slowed down and travel together at the same speed. Because their movement below the surface is restricted the waves topple over and 'break' on the beaches, enclosing pockets of air that burst open forming spray.

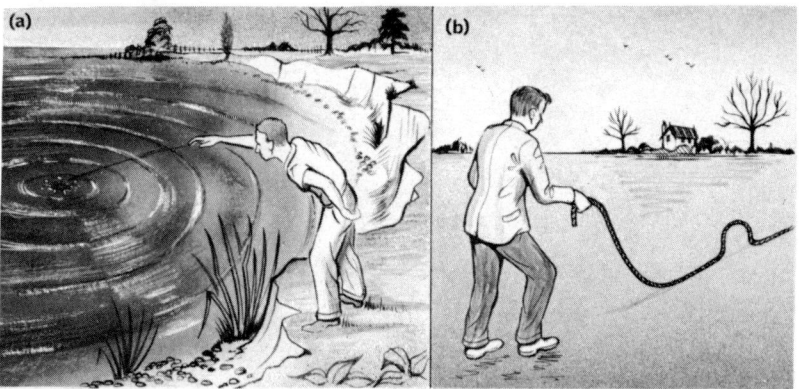

Figure 21.1. (*a*) Waves circling outwards on the surface of a pond. (*b*) A wave travelling along a rope.

Sound also travels in the form of waves from its source and although we can detect these waves if they pass into our ears we cannot see them. Our ears detect them because the sound waves cause the eardrum to vibrate back and forth. If the sound travels through air, the air particles move to and fro (backwards and forwards) along the direction of travel of the waves. This type of wave is known as a *longitudinal* wave; the particles at any one moment are compressed in one position and rarefied in another.

Sound waves can travel through any gas, or liquid, or solid. There must be particles present so that the to and fro vibrations can occur. Sound waves therefore cannot travel through a vacuum.

The speed at which a sound wave travels depends on the density and the elasticity of the material substance through which it travels. The speed of sound in various materials therefore varies greatly. For example, through air at $0°$ C the speed of sound is 332 m s^{-1}, and through iron $5\,000 \text{ m s}^{-1}$, and through water $1\,457 \text{ m s}^{-1}$.

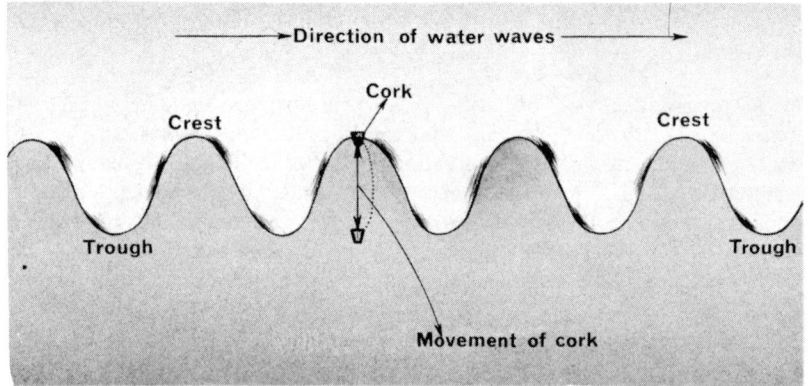

Figure 21.2. A cork floating on a water surface bobs up and down when the wave travels horizontally along the surface. This is a transverse wave.

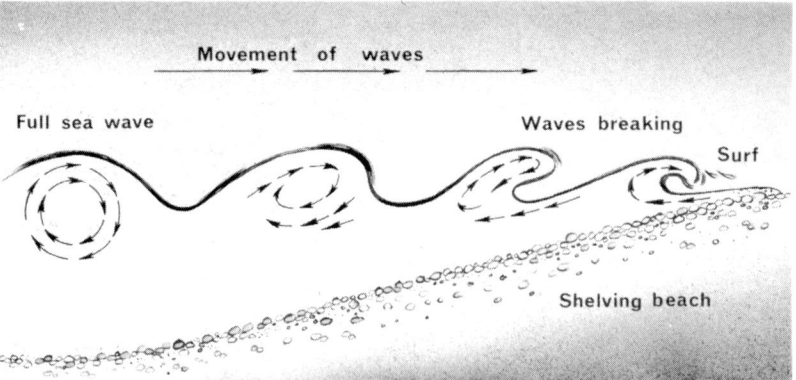

Figure 21.3. A sea wave breaking at the seashore. The movement of the water particles is shown by the arrows.

Earthquake waves. When an earthquake occurs the disturbance sets up waves of various types. There is a longitudinal wave, exactly like a sound wave, that, because of the greater density and elasticity of earth, travels through the earth about five times as fast as a sound wave does through air. There is also a transverse wave, like the shaking wave along a rope, that travels only three times as fast as a sound wave in air. These two waves travel through the earth and can pass through the hard central core of the earth. Another transverse wave travels along the surface of the earth at about twice the speed of sound in air.

Hence an observer with a *seismograph*, that can record earthquake waves in the form of a graph, notices that waves arrive at three different times from the same disturbance. From these readings he can determine the approximate position of the earthquake disturbance.

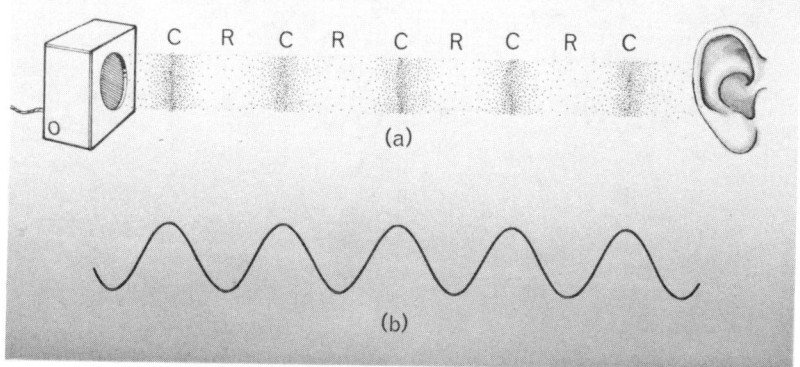

Figure 21.4. Two ways of representing a longitudinal sound wave passing between a loud-speaker and an ear. At any single instant the invisible air particles are compressed (C) or rarefied (R) as shown by the dots in the diagram above (*a*). The same sound waves illustrated by transverse waves is shown in the diagram below (*b*).

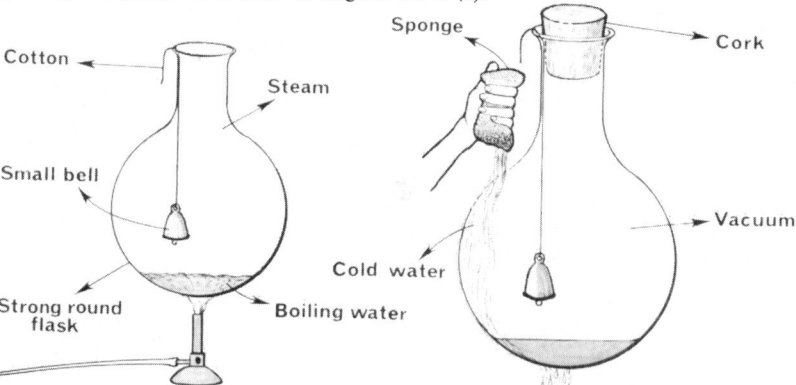

Figure 21.5. This experiment demonstrates that sound waves can travel through air but cannot travel in a vacuum where there is no gas at all. A small bell is hung in a strong round flask in which some water is boiled to drive out the air. The flask is then sealed with a cork and cooled by cold water until the steam has a low pressure because it partially condenses. On shaking, little or no sound is heard from the bell, although a perfect vacuum is not produced.

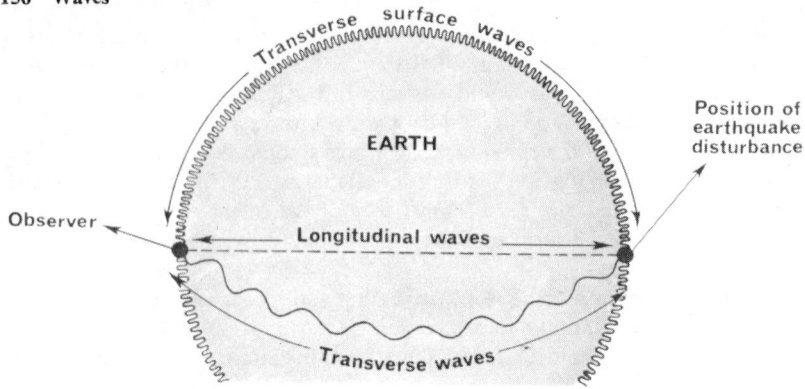

Figure 21.6. This diagram shows how earthquake waves travel through the earth from a disturbance to an observer. The longitudinal waves arrive first and are called primary waves, the transverse waves or secondary waves arrive next, and the transverse surface waves or long waves arrive last.

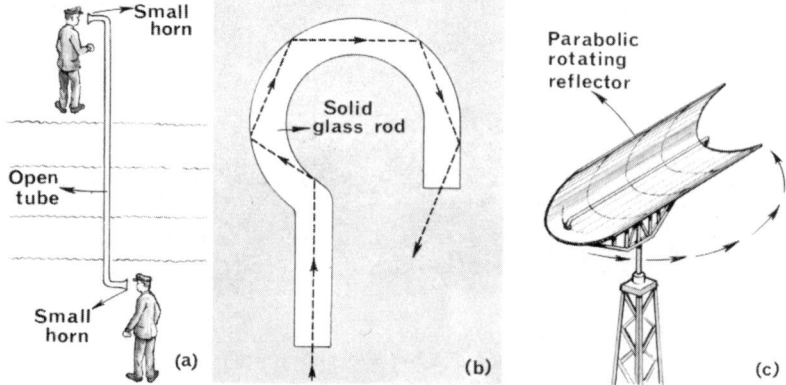

Figure 21.7. The reflection of waves. (*a*) Sound waves along a 'speaking' tube. (*b*) Light waves along a glass rod. (*c*) Radio waves by the curved metal network of a radar transmitter and searcher.

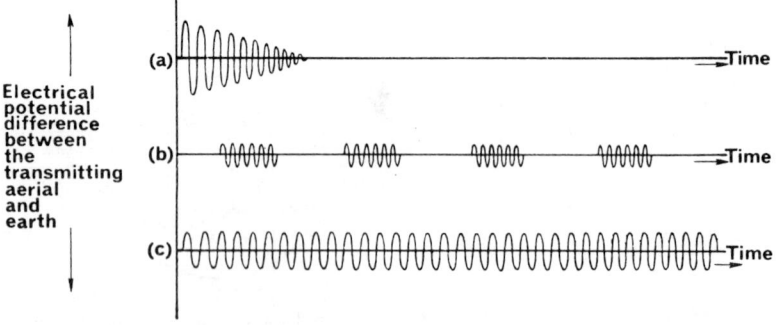

Figure 21.8. This diagram represents the wave forms of (*a*) a pulse of electromagnetic waves produced by an electric spark, (*b*) several short groups of microwaves generated by a radar set, and (*c*) continuous electromagnetic waves used as the 'carrier' waves for radio and television transmissions.

Electromagnetic waves. There is another group of waves that cannot be heard and whose wave form is invisible. We have, however, methods of detecting the effects of these waves. Our eyes can detect *light waves*, our skin can detect *heat waves*, and a radio set can detect *radio waves*. These three kinds of waves belong to a large family of waves called electromagnetic waves. They all travel at the same speed of 3×10^8 m s^{-1} and do not appear to need any substance for their transmission.

Properties of waves. Some waves such as those transmitted along a rope travel in one direction only, others like the water waves on the surface of a pond extend in two dimensions, and others like sound and electromagnetic waves spread out in all three dimensions.

Waves can be reflected, refracted, and stopped altogether. We can 'guide' sound waves along a speaking tube. In the same way we can reflect light along a curved rod of glass. We can 'beam' radio waves by putting the transmitter at the focus of a large metal reflecting 'mirror'.

Waves of the same kind interfere with one another and can be combined to reinforce or cancel one another. They may alternately reinforce and cancel one another and so produce a series of 'beats'. See Figure 23.6.

Waves can also be generated either as a short group together, often called a pulse, or as a continuous series. For example, an electrical spark emits a pulse of electromagnetic waves, and a broadcasting station emits continuous waves by creating a steady oscillation in an electrical circuit. These continuous waves emitted by a broadcasting station are known as carrier waves, and speech or music is imposed on these waves by a process known as modulation.

We have talked of a wave form. What do we mean by this? A cross-section of a number of perfect water waves or any other transverse waves, if we could hold them stationary, would look like those shown in Figure 21.9. This is known as a *sine wave*. One complete movement of a particle of the water up and down and back again to exactly the same relative position on the wave

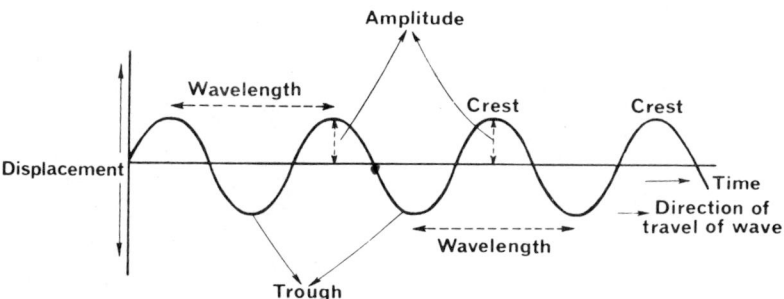

Figure 21.9. A perfect transverse wave. The amplitude and the wavelength are indicated as shown between the arrows.

is known as a *cycle*. The distance travelled by the wave in one complete cycle is the *wavelength*. The number of complete wavelengths that pass any fixed position in a second is known as the *frequency* of the wave and is measured in hertz, Hz (cycles per second).

The speed of the wave is therefore the number of cycles per second multiplied by the wavelength, or:

$$\text{SPEED} = \text{FREQUENCY} \times \text{WAVELENGTH}$$
$$(\text{m s}^{-1}) \qquad (\text{Hz}) \qquad (\text{m})$$

or $$V = f \times \lambda$$

This relationship is true for waves of any kind including the longitudinal sound waves. The wavelength of sound is the distance from one compression of air particles to the next. See Figure 22.4.

The speeds of different kinds of waves are very different and so are the wavelengths and the frequencies. We shall be considering in more detail later the different kinds of waves.

Questions on chapter 21

1. Explain how you would show that sound requires a material medium for its transmission.

2. How does the sound from a tuning fork reach the ear of a person listening to it?

What is the frequency of the note sounded by a siren with 36 holes in the disc when it is rotating at a speed of 420 revolutions per minute? How may the frequency of this note be increased?

3. When man lands upon the moon he will be unable to talk to his neighbour, even if he carries with him his own supply of air for breathing. There will be no such thing as the banging of doors or the clanging of metal. Why is this and how will he be able to communicate by radio with his friends on earth? Will he be able to speak to them and hear their reply?

4. What evidence is there that energy is always required to produce sound?

22. Sound waves

If we examine carefully all the ways that sound can be produced we observe that work is done to produce it in every case. Some force is moved to overcome the resistance of any material substance, which may be the string of a violin, or the air in a clarinet, or the fibres of a piece of paper being torn, or the rubbing of the edge of a glass tumbler by the finger. In other words *energy is always required to produce sound.*

We observe also that *every body that produces sound vibrates.* The skin stretched across a drumhead vibrates, the diaphragm of a loudspeaker vibrates, and the air in an organ pipe vibrates. We can feel all these vibrations with the finger.

A vibrating body is constantly changing its mechanical energy from kinetic energy to potential energy. As the amplitude of the vibrations diminish the mechanical energy is slowly transformed into heat energy and sound. This suggests that *sound is a form of energy.* If this is true then sound can do work. It is possible for us to test this suggestion by a suitable experiment. Scientists work in this way—they are careful to test a suggestion (or hypothesis) by devising an experiment expressly for this purpose, before they state their conclusion (or theory). Open a grand piano and depress the pedal that releases all the wires and so allow them to vibrate freely. Place a small paper rider across one of the wires and sing loudly its note. Stop singing and you will hear that particular wire still sounding that note. At the same time you will observe that the paper rider is disturbed because the wire has been set in vibration by the sound waves emitted by your voice. The sound has given some of its energy

Figure 22.1. A piano wire vibrating in resonance with the sound waves given out by the singer.

159

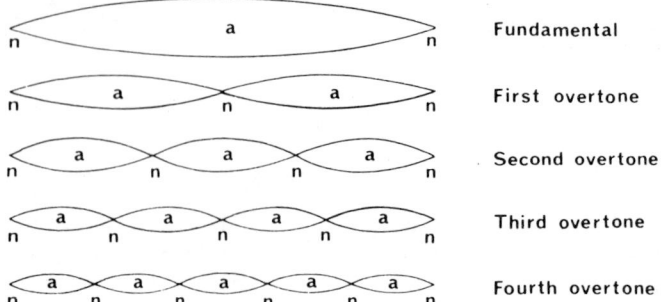

Figure 22.2. This shows how a stretched wire or string can vibrate. The positions marked 'n' are positions where no movement takes place and are called *nodes*. The positions of rapid movement in between the nodes marked 'a' in the diagram are *antinodes*.

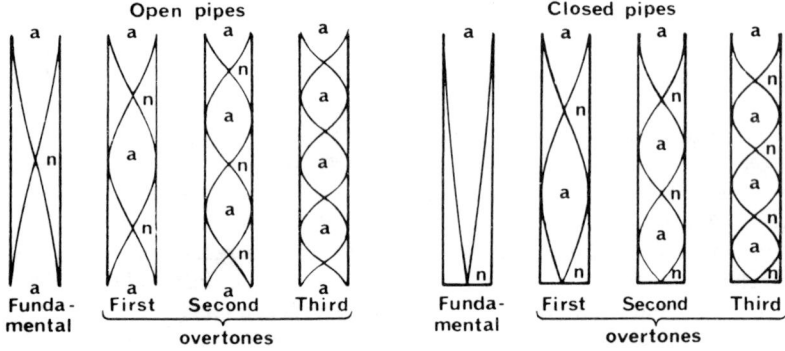

Figure 22.3. The fundamental notes and overtones of open and closed pipes of vibrating air columns. The open end is always an antinode and the closed end a node.

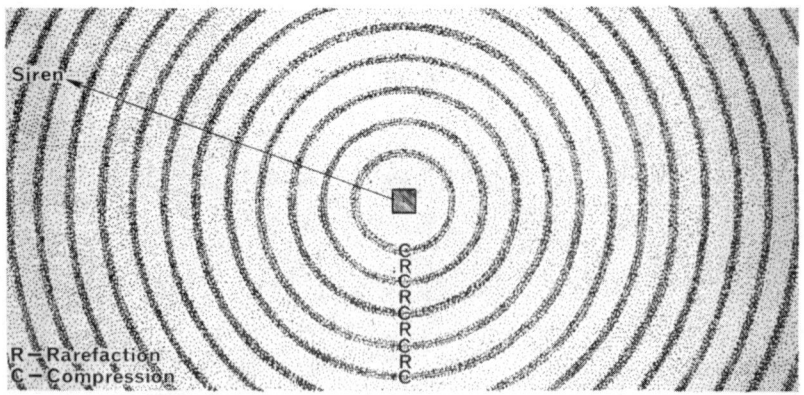

Figure 22.4. A cross-section of the longitudinal sound wave fronts set up around a vibrating siren. This represents the position of the waves at one instant of time for it is a continually expanding pattern of wave fronts.

to that one piano wire. This effect is called resonance and only occurs when the frequency of the note sung to it is exactly the same as the frequency that the wire would itself give out when plucked. From these observations we can conclude that sound is a form of energy.

A piano wire, a violin string, and any kind of stretched wire, string, or rod will vibrate and give out a steady note. It may vibrate because it is in resonance with some strong sound wave or it may be plucked in the centre and thus forced to vibrate. The simple vibration of the whole string gives out its *fundamental* sound note. We call this a *musical* note because it has a steady frequency and a regular wave form. A wave form with an irregular pattern such as would be formed by the sound produced when a tray of crockery is dropped on the floor is known as a *noise*. The frequency of the fundamental note given out by a string depends on its length, tension, and mass. Look carefully at the wires in a piano to see how the length, tension, and mass of the wires are varied to produce the different notes.

A string may also vibrate in smaller segments than the whole length if plucked in other positions, not the centre, each segment being an exact fraction of the whole length. The musical note then emitted by the string will contain the original 'fundamental' note and also higher notes called *overtones*.

An organ pipe, a whistle, a recorder, a clarinet, and other wind instruments also contain columns of air that can vibrate. There are vibrations produced by the whole length of the air column that give the fundamental musical note, and vibrations produced by fractions of the whole that give the overtones. There are nodes and antinodes as in the case of the vibrating string. The actual note given by a wind instrument depends on the length of the vibrating air column in the tube and this is controlled by the position of the 'stop' un-covered. Some of the tubes are open at both ends and others are open at only one end. Different instruments have different methods of starting the vibra-tions; some use a reed; others rely on the lips or a sharp edge.

It is more difficult to describe how longitudinal sound waves travel than how transverse waves travel. Perhaps you have seen waves travelling across a large expanse of standing corn and watched the ears of corn vibrating to and fro, or perhaps you have seen and heard the jerk as a long line of railway wagons is started by a sudden pull from the engine in front. The jerk or the 'to-and-fro movement' passes from wagon to wagon along the line of wagons as the 'wave' travels longitudinally from one end to the other.

A siren emitting continuous sound waves sets up a pattern of compressed and rarefied air in concentric spheres that expand outwards all the time. At any particular moment these spheres if cut through and drawn on paper would show a cross-section of curved wave fronts.

The *frequencies* and the *wavelengths* of various sounds are related by the general equation that holds for all wave movements.

$$V = f \times \lambda.$$

We have noticed the delay between seeing a flash of lightning and hearing the associated clap of thunder, between the sight of steam escaping from a ship's whistle and the arrival of the sound it made. In both these cases the light from the disturbance travels much more rapidly than the associated sound wave. The time taken by the light to reach us from such a disturbance can be neglected because light travels so rapidly (3×10^8 m s^{-1}). Thus it is an easy matter to calculate the speed of sound (V)—we only have to know how far the disturbance is away and how long the sound takes to arrive. A gun is fired in some distant position and an observer notes the time that elapses between the arrival of the flash and the sound. In case a wind is blowing along the track another measurement is also taken in the opposite direction and the average speed is calculated. The speed is about 332 m s^{-1} at $0°$ C.

Each *musical note* has a certain frequency and therefore a wavelength that remains constant for that particular note in air. These properties identify the note. A note having a frequency of 256 vibrations per second (256 Hz) is called *middle C*. We often use the word *pitch* when talking of a musical note. When the frequency of one note is twice that of another we say that the difference in pitch is an *octave*. Figure 22.6 shows how other simple ratios of frequencies between 1 and 2 produce notes that are related to one another in what musicians call the *major diatonic scale*.

Starting with the key note of middle C the scale of C can be represented by eight notes the ratio of whose frequencies is given by the numbers 24, 27, 30, 32, 36, 40, 45, and 48. This *scientific* scale does not prove suitable for use when tuning a piano. Thus the tuner learns to make slight adjustments so that the musician can play in any key without the music sounding rather queer in some keys.

The *human ear* is very sensitive and can recognize notes whose frequencies vary from about 15 Hz to very nearly 20 000 Hz. Compare this range of frequencies with those of some musical instruments described in Figure 22.8.

Many animals can hear sounds of even greater frequencies than can humans.

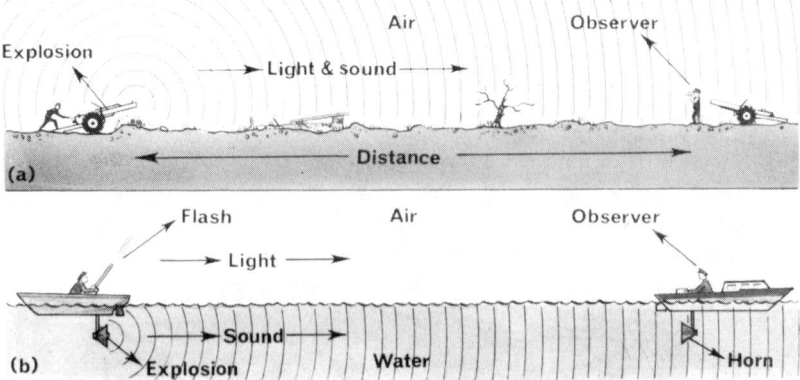

Figure 22.5. The speed of sound is measured (*a*) in air, and (*b*) in water. The gun fired in water sets off a flash at the same time in the air above the boat, and the observer sees the flash and hears the explosion through a collecting horn in the water.

Bats cannot see well, and although they usually fly at night, they succeed in avoiding objects in flight by sending out high frequency sound waves and use the echoes that bounce back to them from near objects to guide them. Some young people can just hear the shrill note of a bat. There is a whistle that when blown sends out a note of such a high frequency that it cannot be heard by human beings. But dogs can hear it and they can be trained to respond to its message.

A musical note may change in *quality*. See Figure 22.9. A note played on a violin is easily recognized as different from one played on a horn even if they have the same frequency. This difference is explained by the difference in wave form which may be partly due to the presence of overtones imposed on the main frequency. These overtones result from the manner in which the sounding body is set in motion.

The *amplitude* of the wave form produced by a vibrating body is one factor that determines the *loudness* or *volume* of the note it emits. See Figure 22.9. Loudness also depends on the surface area of the vibrating body.

The intensity of sound to which the human ear can respond covers a very considerable range. The lower limit of audible sound is known as the *threshold of hearing*. Ordinary conversation is a million times stronger than this limit. The upper limit is known as the *threshold of feeling* and at this stage the sound intensity is again a million times stronger than it is for ordinary conversation.

The relative sound intensities can be expressed in powers of ten: $1, 10^1, 10^2, 10^3, 10^4, \ldots 10^{12}$, but it is simpler to consider the indices themselves as a measure of the *sound level* and call these *bels*. A difference of sound level of one-tenth of a bel is a *decibel*.

The following list gives very approximate values of the intensities of various sources of sound expressed in bels from the threshold of hearing (0 bels) to the threshold of feeling (12 bels).

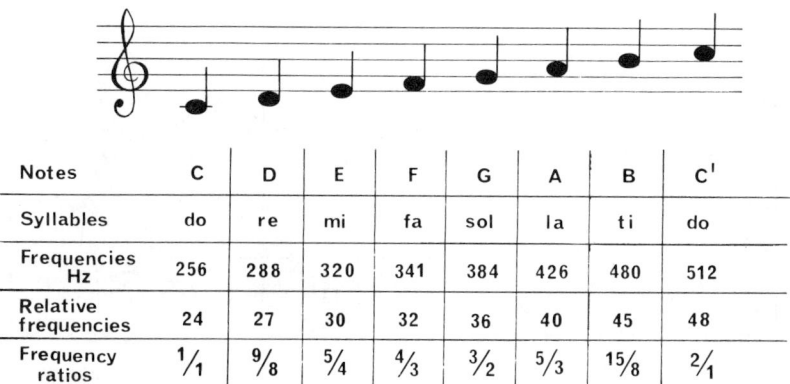

Notes	C	D	E	F	G	A	B	C^1
Syllables	do	re	mi	fa	sol	la	ti	do
Frequencies Hz	256	288	320	341	384	426	480	512
Relative frequencies	24	27	30	32	36	40	45	48
Frequency ratios	$\frac{1}{1}$	$\frac{9}{8}$	$\frac{5}{4}$	$\frac{4}{3}$	$\frac{3}{2}$	$\frac{5}{3}$	$\frac{15}{8}$	$\frac{2}{1}$

Figure 22.6. The relationship between the notes of the major diatonic scale of C.

Bels	Bels
0 Threshold of hearing	7 Loudspeaker
1 Human breathing	8 City traffic
2 Quiet countryside	9 Aircraft at take-off
3 Burning bonfire	10 Moving underground train
4 Average house noise	11 Peal of thunder
5 Motor-car engine	12 Threshold of feeling
6 Ordinary conversation	

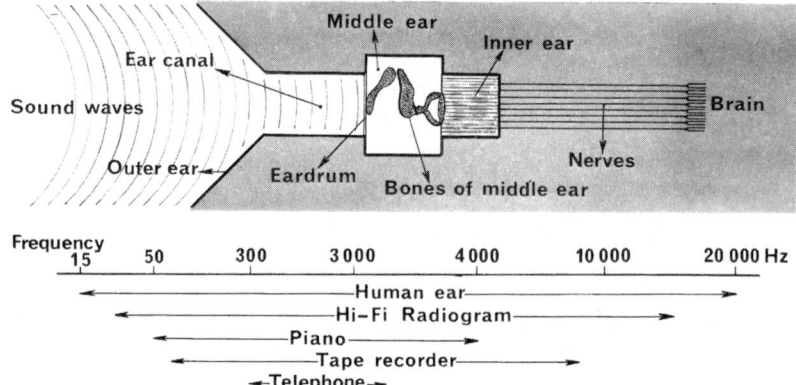

Figure 22.7. This diagram illustrates how the human ear hears. The sound waves strike the eardrum and the vibrations formed are passed through the middle ear to the inner ear where nerves convey the messages to the brain.

Figure 22.8. The frequency range of the human ear and various instruments.

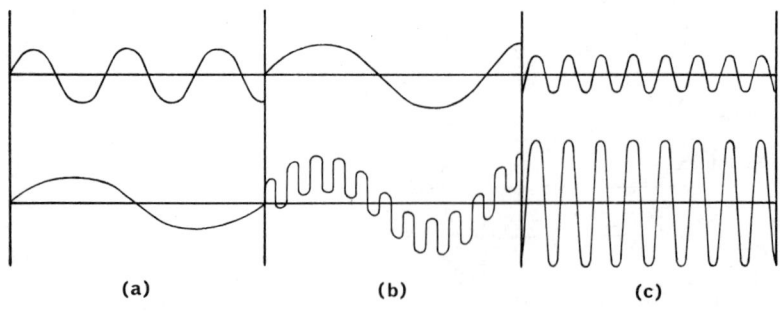

Figure 22.9. Pairs of transverse wave patterns representing pairs of longitudinal sound waves of different (*a*) pitch, (*b*) quality, and (*c*) loudness.

Questions on chapter 22

1. Sounds are made by vibrating bodies that in turn cause the air to vibrate. How would you alter the vibration to make the sound (a) louder, and (b) give a higher note?

2. A string vibrates 550 times a second when the velocity of sound in air is 335.5 m s^{-1}. What is the wavelength in air of the note sounded?

3. Describe an experiment to show that sound needs a material medium in which to travel, while light does not. In what other important respect does the transmission of sound differ from that of light?

4. Describe briefly how you would make a rough determination of the velocity of sound in air.

5. Describe an experiment to show that sound will not travel in a vacuum.

6. What determines the pitch of a musical note? Why may a violin which has been tuned in the open air sound 'flat' when played in a warm room?

7. What is meant by the frequency of a sound wave? How is this quantity connected with the velocity of sound?

8. Mention three common observations which illustrate that sound travels more slowly than light.

9. What do you understand by (a) the pitch, and (b) the loudness, of a note? How can you raise the pitch of a note given out by a stretched string?

10. Calculate the velocity of the sound waves sent out in air by a string if the note heard has a frequency of 275 hertz and the wavelength is found to be 1.2 metres.

11. What must you do to raise the pitch of the music emitted by a gramophone? What does this do to the quality and to the harmony of the music? Explain your answers.

12. How do you account for the fact that the front rows and the rear rows of a long column of soldiers on the march are slightly out of step when led by a band in front?

13. Why do the bass, baritone, alto, and soprano saxophones vary so much in size?

14. Use diagrams to explain why notes of the same pitch sound different when played on different instruments. Why, for example, is the note middle C on a piano very different from the note middle C played on the violin?

23. The nature of waves

Reflection. Waves are reflected when they meet some obstruction in their path through which they do not easily travel. Circular water waves formed in a swimming bath bounce off the retaining walls in regular patterns, and sea waves can be observed to be reflected away from a stone harbour wall. A rope fastened at one end to a wall and held at the other end in your hand can be made to reflect back along itself a wave formed by a single rapid shake. Examine again Figure 21.1 and compare it with Figure 23.1.

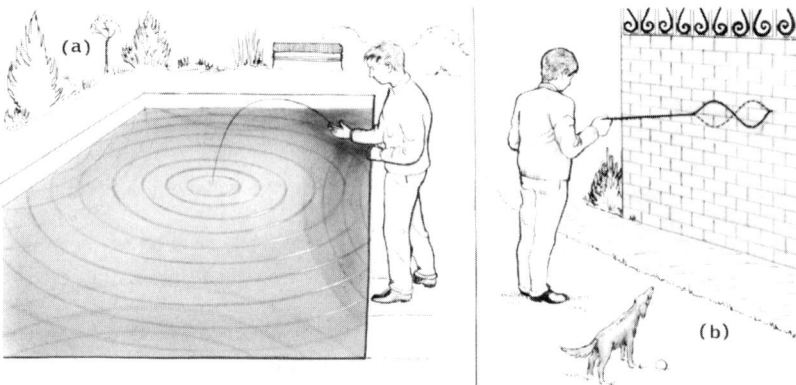

Figure 23.1. The reflection of (a) transverse water waves, and (b) a wave travelling along a thin rope.

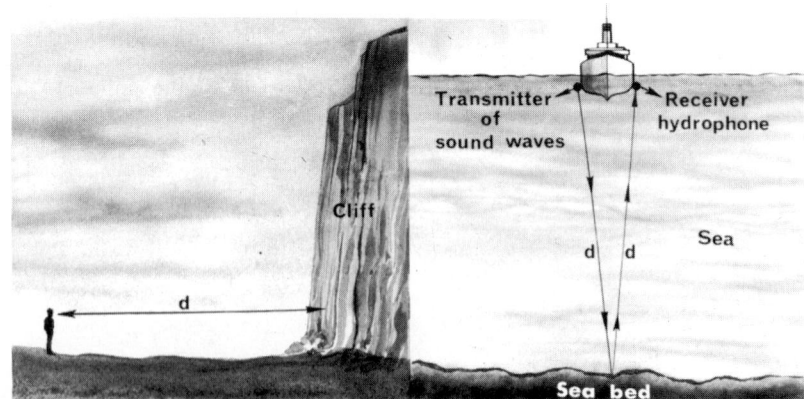

Figure 23.2. The reflection of sound waves. The distance (d) of the reflecting surface can be calculated if you know the speed of sound in the air or water medium (v) and the time the echo takes to return (t) because $d = \frac{1}{2}vt$.

166

Longitudinal sound waves are reflected as echoes from cliffs and the walls of buildings. Climbers when lost during misty weather in mountain valleys can find their way through them by making use of echoes. Captains of ships can determine the depth of water below them or the depth of a shoal of fish by using an underwater device emitting certain sound waves and receiving their echoes. The sound waves used are sometimes supersonic, that is, they have a frequency higher than the highest frequency that a human ear can detect.

Light waves are reflected from polished metal surfaces or silvered glass mirrors. These surfaces can be of any shape but if they are flat or regularly curved they form clear and real or virtual images. See chapter 2.

Radio waves, like other waves, are also reflected and, in fact, the principle of an echo is used in the working of a *radar* set as will be explained more fully in chapter 32. Short bursts of microwaves are transmitted, and data from the reflected waves enable the controller to determine the direction and distance away of the reflecting conductor. In this way aircraft can be followed in flight, and sailors can navigate their ships during foggy weather with less danger of collision.

Refraction. At the boundary between one medium and another, waves passing through change their direction unless they happen to strike the boundary normally. This bending is called refraction.

Water waves change direction as the waves cross over a shelf into shallow water.

On a hot day sound waves bend as they are refracted in passing from colder air to the warmer air near the surface of the earth. The bending is gradual as the air changes its temperature gradually. A person's voice does not carry along the ground as far in the daytime as it does at night as can be seen in Figure 23.4.

Light waves are refracted or bent as they pass from a rarer medium to a denser medium or vice versa. It is for this reason that a lens can bend light rays to or from a focus. See chapter 3.

Radio waves are bent as they pass into a layer of electrified air about 240 km above the surface of the earth. See Figure 32.1.

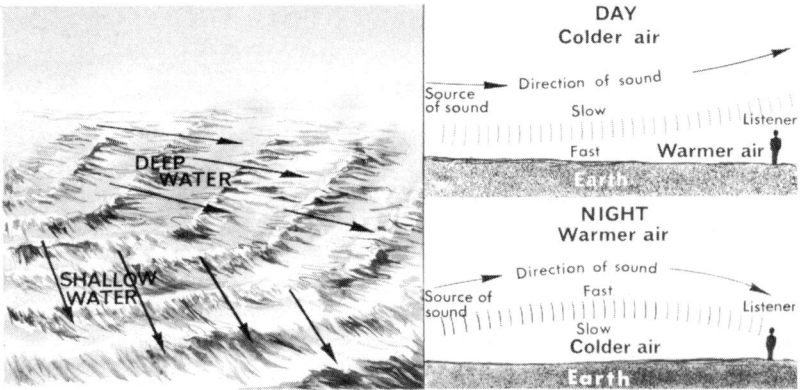

Figure 23.3. This diagram shows how water waves bend when they cross from deep to shallow water. The waves travel faster in deep water than in shallow water. The direction of travel is shown by arrows.

Figure 23.4. These diagrams show how a very narrow beam of sound waves bends by day and night. There are many other beams besides the two selected and they are all bent in a similar way. Sound travels faster in warm air than in cold air.

M

Interference. The waves generated by two separate sources can interfere with one another if they both arrive at the same point together. This is known as interference. It may happen that each crest of one transverse wave coincides precisely with each trough of the other transverse wave. If they have equal amplitudes they cancel one another and the position is left undisturbed. This happens in water waves when a reflection produces a wave that encounters the original or another similar wave. The result of the meeting is a calm undisturbed area. If two crests or two troughs coincide the disturbance is increased and perhaps doubled.

It is easier to understand the effects of two waves meeting one another if we draw them as two transverse waves even though they may be longitudinal waves.

The regular loud, soft, loud, soft, loud, soft . . . sounds heard when two almost identical notes are struck are known as *beats*. Beats can be heard when two identical engines are running at slightly different speeds. A piston-driven twin-engined aircraft provides such a possibility and one often hears the throbbing sound produced by these two engines. Musicians listen for beats when they adjust the strings of their cellos or violins to obtain equal frequencies. With a standard tuning fork a piano tuner is able to gauge the frequency of a particular wire by listening to the sound beats produced as the two are made to vibrate at the same time.

Patterns of the interference between the waves produced by two sources can be seen easily when two equal stones are thrown into a pond close to one another at the same time. Two circular wave patterns spread out and interfere along quite well defined paths. In some places the waves aid one another and in other places they oppose one another.

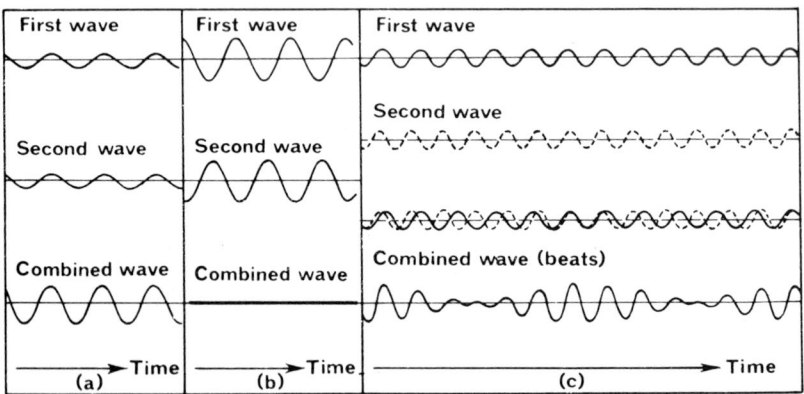

Figure 23.5. The combination of two transverse waves. (*a*) Two waves of the same frequency and exactly in step (in phase) with one another aid each other. (*b*) Two waves of the same frequency and exactly out of phase oppose one another. (*c*) Two waves of slightly different frequencies, that are alternately in phase and then out of phase, produce *beats* because they alternately aid and oppose one another.

The same interference patterns in sound occur round a tuning fork. Strike a tuning fork and hold it close to your ear and then turn it round slowly. In certain positions the sound is loud and in others it is very quiet.

It is difficult to see the interference patterns in the case of light because the wavelengths are so short. Oil films on water are coloured because light waves, that are reflected from the top and bottom surfaces of the oil, interfere with one another. The light from a distant source, passing through the fine mesh of an umbrella or a handkerchief, forms regular light and dark patterns due to the interference of the light coming through the many different small holes of the material.

Questions on chapter 23

1. A battleship fired a shot and the echo from distant cliffs was heard aboard four seconds later. How far away were the cliffs? Explain, with a diagram, how this echo was formed. Assume that the speed of sound in air is 332 m s^{-1}.

2. Draw diagrams to illustrate how sound can be used to (a) prospect for oil layers in the earth, (b) measure the depth of the seas.

3. Why is it more difficult for a mountaineer to make his friend hear when he is a little distance away when they are on top of a mountain than when they are at its base?

4. In a large auditorium we make music much louder but more blurred by removing the sound absorbing boards attached to the walls and ceiling. Explain this statement.

5. Why is it easier to talk in a large auditorium filled with people than in the same auditorium when it is empty?

6. Why do people who are a long way from a speaker cup their hands around their ears when listening?

7. Describe an experiment that you could perform to illustrate the formation of beats if you were given two identical loud-ticking mechanical clocks.

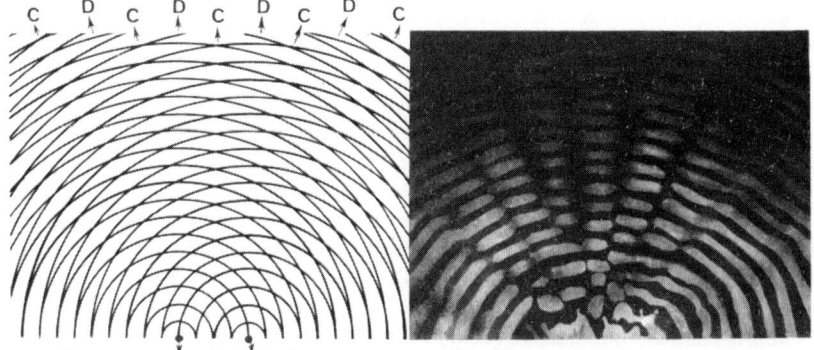

Two sources of disturbance   Resultant water wave pattern

Figure 23.6. Interference patterns of water waves on the surface of a pond. Lines marked C represent the positions of constructive interference where the waves build up and aid one another. Lines marked D represent positions of destructive interference where the waves cancel one another.

24. Electromagnetic waves

Michael Faraday, an English physicist and chemist who became well known for his work on electromagnetic induction, assumed the existence of some medium capable of transmitting electric and magnetic forces, but he had no idea of its nature.

It was about 1865 that the Scottish physicist James Clerk Maxwell, basing his argument upon purely mathematical reasoning, proposed that light was a form of electromagnetic energy. He was able to show that light and electromagnetic waves were both transverse waves and travelled at the same speed. Maxwell also predicted that electromagnetic waves would be reflected by metallic objects, and that they would bend upon entering certain obstacles.

This work of Maxwell led to the discovery of certain electromagnetic waves by the German physicist Heinrich Hertz who was actually able to send and receive them in his laboratory. He called them radio waves and these were the first ever sent by man. Less than twenty years later Marconi transmitted and received these waves across the Atlantic Ocean.

It was shown in the meantime that radiant heat could be reflected and refracted in the same way as light and electromagnetic waves. Radiant heat also travelled at the same speed as light and electromagnetic waves.

The idea of an electromagnetic spectrum embracing all the kinds of waves known at that time was then put forward. This spectrum includes radio waves, radiant heat waves, and light waves. The general acceptance of the electromagnetic spectrum as shown by the table on page 171 enables scientists to explain many useful features of the electromagnetic waves.

Scientists predicted other waves of different wavelengths then unknown, and slowly filled in the whole range of the spectrum as it is known today. Thus much information became available about the nature of the various waves and how they can be produced, detected, and used to the benefit of mankind. It is probable that there are waves within and beyond the present electromagnetic spectrum about which more will be discovered in the future.

Scientists admit that at present they do not know the real nature of all these radiations. They know it requires energy to produce them and that they are thus some form of energy. It also appears that matter is a form of energy. In fact, matter and energy are evidently interchangeable, and Einstein has been able to calculate the tremendous amount of energy available when matter is converted into energy.

It is suspected that matter is very slowly converted into energy in the sun and other stars and radiated as heat and light waves. The sun has been emitting very large quantities of energy for long periods of time and is apparently losing

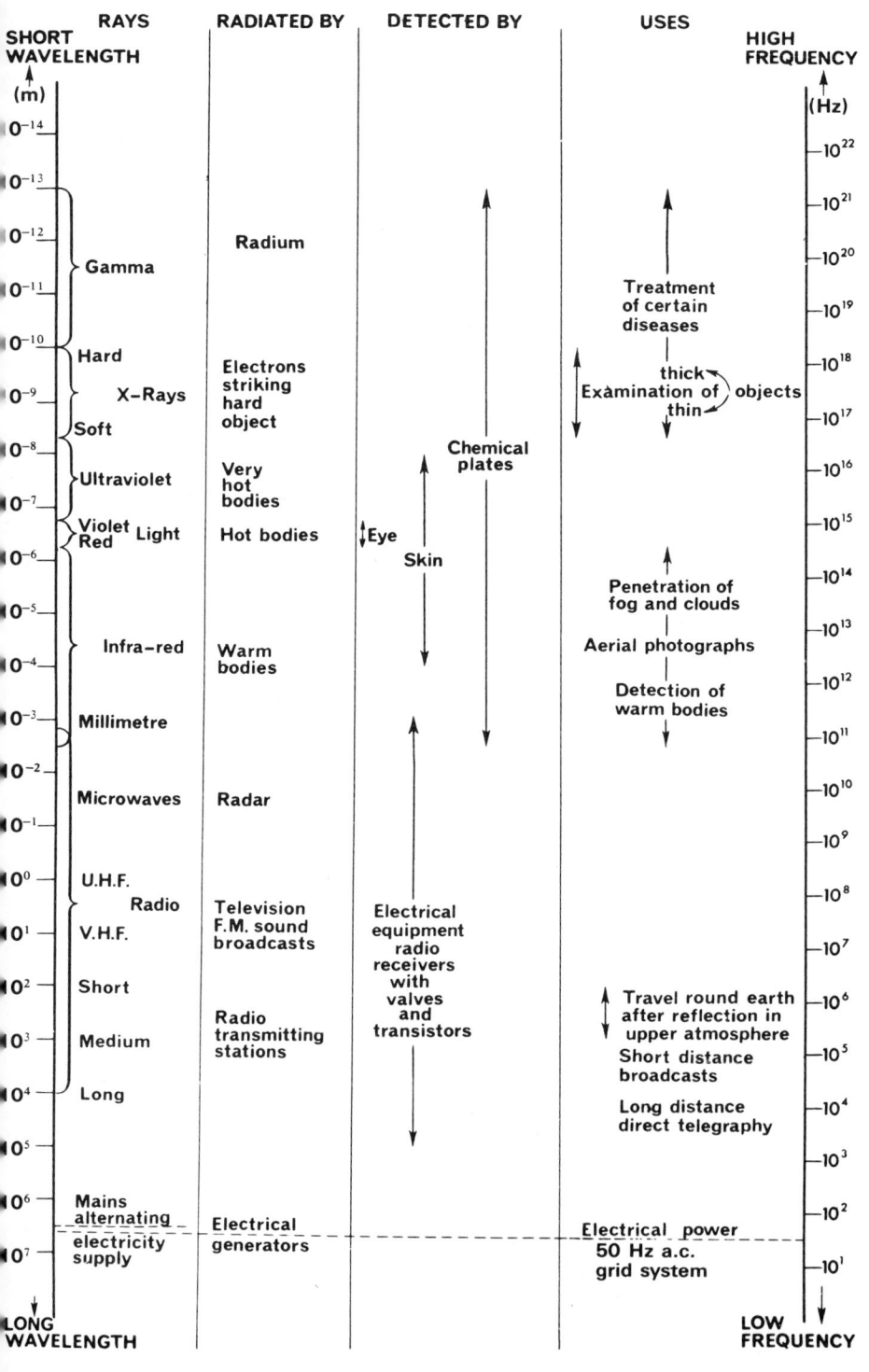

very little matter. This transformation of matter into energy could continue for many millions of years more.

The matter and energy relationship also enables us to calculate how much atomic fuel is required to supply the energy for the new electricity generating stations and for the nuclear-powered ocean liners of the future. It is because of this relationship that a small quantity of matter in an atomic bomb can create such a tremendous amount of energy in its explosion.

The wave equation discussed in chapter 21 holds for all the radiations of the electromagnetic spectrum. The radiations all travel through space at the high speed of 3×10^8 metres per second. The ranges of the frequencies and the wavelengths of the radiations are given on the right hand and left hand sides of the table on page 171. The frequency and wavelength of any particular radiation are related to one another by the equation:

$$\text{SPEED} = \text{FREQUENCY} \times \text{WAVELENGTH}$$
$$(3 \times 10^8 \text{ m s}^{-1})$$

Various instruments must be used to detect many of the radiations in the electromagnetic spectrum. Radio waves of all wavelengths are received, detected, amplified, and possibly converted to sound waves that we can hear by radio receivers.

The microwaves used in radar and relay stations can be employed only where there are no obstructions in the direct line between the transmitter and the receiver. Television programmes are passed by a chain of relay stations from hill top to hill top on microwaves.

The infra-red or heat waves are detected by our skin and are quite invisible. They can be felt near the heating bar of an electric fire when it is first switched on long before the bar appears to glow. Infra-red waves act on a photographic plate and thus they can be used for photography. These waves penetrate fog and clouds fairly easily and therefore are used to photograph distant scenes in misty or cloudy weather. Hot bodies not emitting visible light can still be photographed in the dark with infra-red waves.

Our eyes respond to a small range of radiations known as the visible or light waves. The different wavelengths represent different colours in the light spectrum. The whole band of the light spectrum occupies only a very small part of the electromagnetic spectrum.

Ultra-violet waves, plentiful in sunshine, cause sunburn of the skin and also produce vitamin D in the body. They cause some living tissues, like our teeth and finger nails, to glow or fluoresce. Many other substances glow brilliantly when the invisible ultra-violet waves fall on them. Some of these substances are painted on the inside of fluorescent electric lamps; there they receive strong ultra-violet waves from the electrical discharge in the lamp and convert these waves into waves that are visible to the eye. Other similar substances are mixed with paints and dyes, and these are often used for advertisements because they 'glow' brilliant colours (usually red, yellow, or green) when they convert the ultra-violet waves of the ordinary daylight that falls on them.

X-rays and gamma rays can be detected by their chemical action on specially prepared photographic plates. These rays destroy skin and other cell tissues that are exposed to them for more than a few seconds. They pass fairly freely through human flesh but are mostly stopped by bones.

The table on Page 171 gives the main properties and uses of the radiations in the electromagnetic spectrum. The list is incomplete because scientists are continually adding to their knowledge of these radiations.

Questions on chapter 24

1. What is the essential difference between all the kinds of waves in what is known as the electromagnetic spectrum?

2. How many different kinds of waves that can be reflected do you know in the electromagnetic spectrum?

3. What have all the kinds of waves in the electromagnetic spectrum in common?

25. Magnetism

A tale is told that long ago a shepherd sat down on a rock and rested his crook against it. When he got up he found to his surprise that the iron crook stuck to the rock! The shepherd was sitting on *loadstone*, a dark brown iron ore that possesses the peculiar properties not only of attracting iron, but also, when a suitable piece of it is freely suspended, of always coming to rest pointing in a north-south direction. The Chinese from very early times used this stone to help them on their travels over land and sea. They would hang a piece of it up so that it could 'lead' them and hence it got its name 'loadstone' or 'lodestone'.

It is believed that the first specimens of loadstone were found over 2 000 years ago near Magnesia, an ancient city in Asia Minor, and this is the reason that loadstone is also called 'magnetite'. The word 'magnet', as you can guess, is also derived from the name of this ancient city.

A steel magnet, like a piece of loadstone, when freely suspended, comes to rest pointing to the north and south. The end which points north is called the *north-seeking pole* or *N. pole* and is often painted red. The other end, pointing to the south, is called the *south-seeking pole* or *S. pole* and is often painted black. The simple compass of today, consisting of a steel magnet pivoted on a fine steel point is often used in the same way as the Chinese used their pieces of loadstone for 'leading' travellers during their journeys.

The poles of a magnet are not situated at the extremities of the magnet but at some small distance away from them inside the magnet. We can estimate the actual positions of the poles by observing how small pins or iron filings are

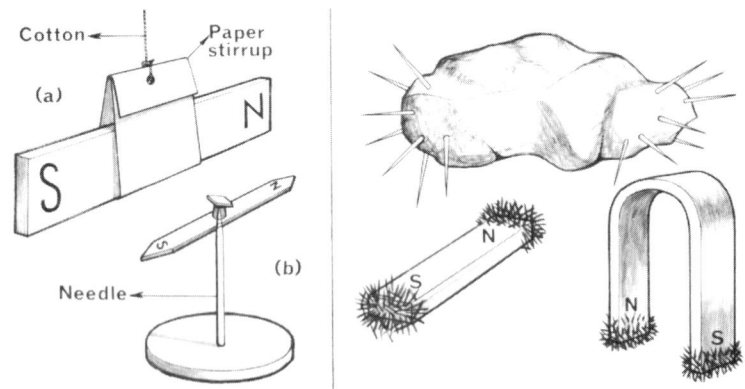

Figure 25.1. Two methods of giving a long steel magnet freedom of movement horizontally so that it can come to rest where it wishes: (*a*) suspended in a paper stirrup, and (*b*) mounted and balanced on a fine needle point.

Figure 25.2. A piece of loadstone, a bar magnet, and a horseshoe magnet attracting pins and iron filings. Where are the poles of the magnets in each case?

174

attracted to a magnet at various places on its surface.

A magnetic field is the region around a magnet where the force of attraction or repulsion due to the magnet can be detected. Two simple ways of *observing the shape* of a magnetic field are by sprinkling iron filings onto a white card placed over the magnets (Figure 25.3), and by plotting the field with the aid of small sensitive balanced magnetic needles (Figure 25.5).

Attraction and repulsion. It is possible to perform many simple experiments with two magnets to observe that similar or like poles repel and unlike poles attract one another. An unmagnetized piece of iron is always attracted by the N. pole and also by the S. pole of a magnet. It is never repelled. Hence a certain way of knowing that a piece of iron is magnetized is to determine if it can be repelled at one end by one of the poles of a magnet. If there is no repulsion then the piece of iron is not a magnet.

Every magnet has a N. pole and a S. pole and we have never found any other kind of pole. It is impossible to separate a single magnetic pole from the other of its pair. When we examine magnetic fields we always find that there are complete *lines of force* running from a N. pole to a S. pole. The iron filings set themselves along these lines of force and we define the direction of a line of force as the direction that a N. pole will try to move along it.

Each portion of a magnet forms by itself a complete magnet with a N. pole and a S. pole. We can observe this by breaking a long thin magnetized needle into two pieces. Each section will be found to be a complete magnet. This process can be continued indefinitely and leads us to the theory that each molecule of iron is a small magnet. The theory then supposes that if all the molecular magnets point in one direction the piece of iron of which they form part is magnetized, but if they are mixed around haphazardly the iron is un-magnetized. This can be demonstrated by nearly filling a test-tube with iron filings and stroking the tube with a magnet. Observe the movement of the

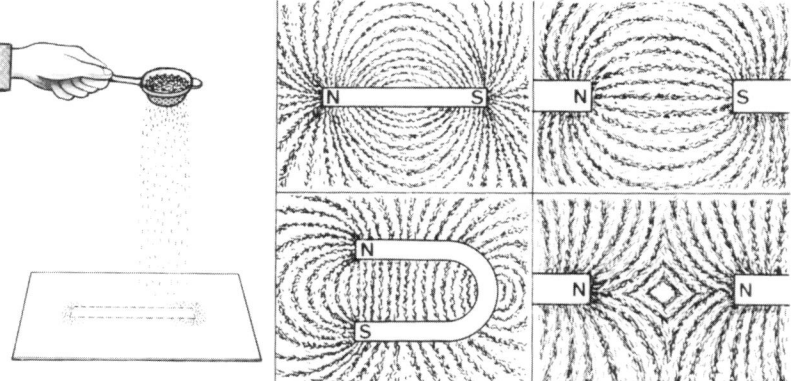

Figure 25.3. The magnetic field in the neighbourhood of various magnets can be shown by sprinkling iron filings onto a sheet of white card placed over the magnets.

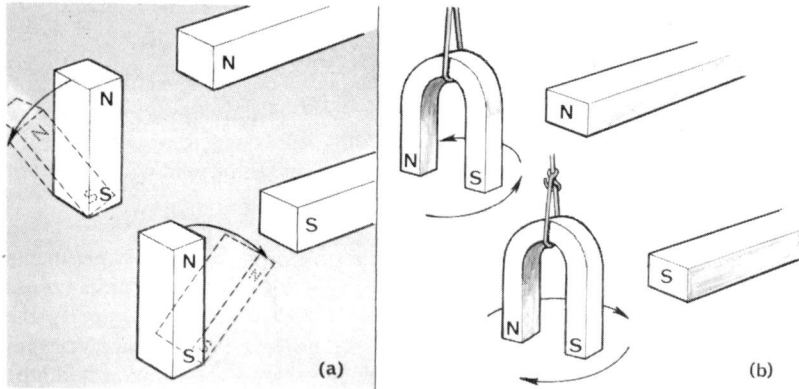

Figure 25.4. The attraction and repulsion of magnetic poles. Depending on the kind of pole presented (*a*) the bar magnet standing on its end falls over one way or the other, and (*b*) the horseshoe magnet twists itself in one direction or the other.

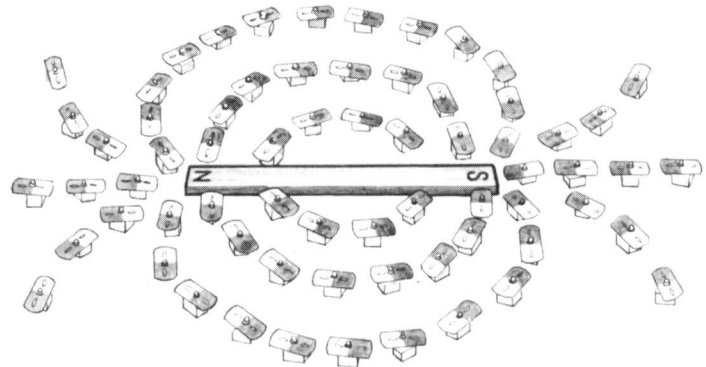

Figure 25.5. Magnetized razor blades pivoted on needles mounted in corks are shown here being used to determine the position and the direction of the magnetic lines of force around a bar magnet.

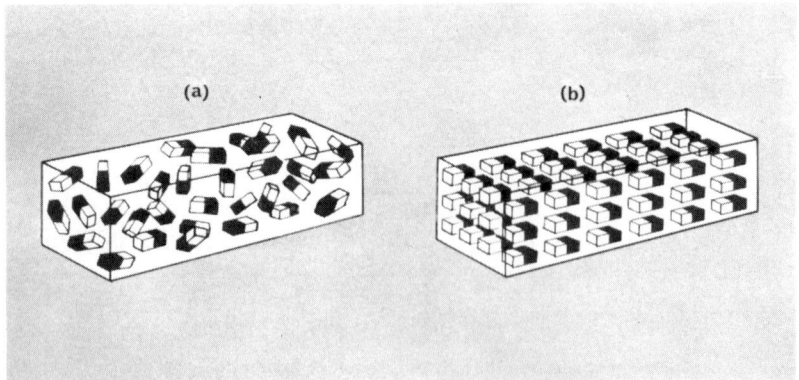

Figure 25.6. Arrangement of the molecular magnets in a bar of (*a*) unmagnetized iron, and (*b*) magnetized iron.

filings and test the magnetic field around the tube. Shake the test-tube, observe the filings, and test again.

Induced magnetism. If a piece of unmagnetized iron is merely held near a magnet it becomes magnetized by a process called *magnetic induction*. By stroking the unmagnetized iron with the magnet the induced magnetism produced is made stronger. In both cases the original magnet does not lose any of its strength.

The magnetic field of the earth. We have observed that freely suspended magnets always set themselves at any particular place in the same horizontal direction along a line running from the north to the south. This line is called the *magnetic meridian* and it varies in direction considerably over the surface of the earth and also from year to year. The angle between the magnetic and the geographical meridians at any point is known as the *angle of declination* or the *variation* of the compass at that point. Travellers used to rely on a carefully balanced magnet in order to determine the direction they were moving. Today radio bearings are easier and more reliable to use for this purpose, and they have the advantage that there is no variation to be taken into account.

We have so far considered only magnetic needles that can swing in a horizontal plane. A magnetic needle mounted so that it can rotate in a vertical plane is found to 'dip' at most places on the surface of the earth. The *angle of dip* formed between the horizontal and the axis of the magnetic needle also helped travellers in former days to find out precisely where they were because they had maps showing these angles at all places on the surface of the earth.

We talk about the magnetic field of the earth as if the earth contained a large magnet whose poles are not far removed from the geographical poles of the earth. Certainly when we look at Figure 25.9(*b*) the earth appears to have two magnetic poles, but, of course, the interior of the earth is too hot for there to be a magnet. We do not know the origin of the magnetic field of the earth,

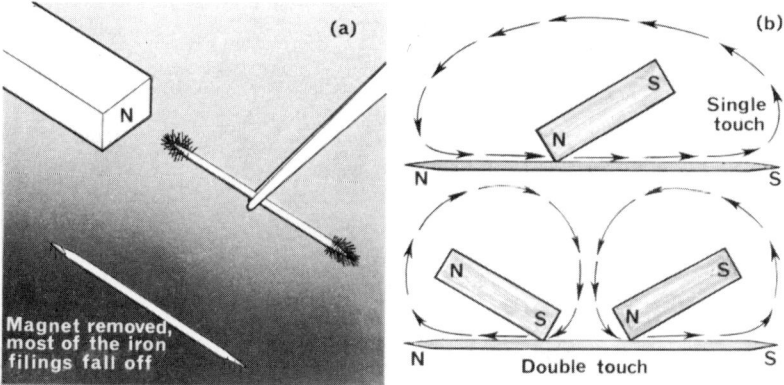

Figure 25.7. The process of magnetic induction. An unmagnetized iron needle is (*a*) placed near a magnet, and (*b*) stroked by one or two magnets.

we only know its shape and strength. Why do you think we have called the magnetic pole near to the geographical north pole a north magnetic pole when it is really the south pole of the imaginary large magnet inside the earth? The answer can be found if you will think carefully about the meaning of a N. pole and a S. pole as described near the beginning of this chapter. The north magnetic pole is situated near the Queen Elizabeth Islands in Northern Canada, and the south magnetic pole is just inside Antarctica and has nearly the same longitude as Tasmania. The positions of these two magnetic poles vary slightly from year to year.

Figure 25.8. A prismatic compass is used by this walker to find his bearings in unknown country. His readings give magnetic bearings.

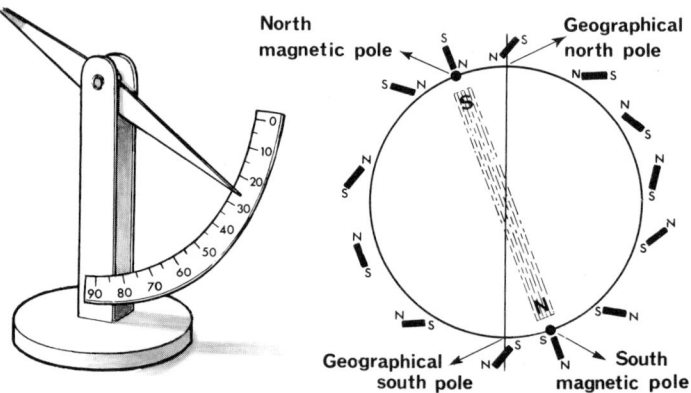

Figure 25.9. (*a*) A simple 'dip' needle. (*b*) The approximate positions taken up by a 'dip' needle around the surface of the earth.

Questions on chapter 25

1. Explain the differences between a temporary and a permanent magnet. Give examples of the uses of both types of magnet.

2. If you were given two bar magnets with the poles of one magnet un-labelled and of the other labelled, how would you discover with these two magnets alone the polarity of the unlabelled magnet?

3. Describe how you could make a magnet from a piece of unmagnetized iron if you had also two bar magnets. Draw diagrams to show the different methods and indicate the polarity of the magnets used and the magnet created.

4. What is meant by the molecular theory of magnetization? Describe two simple experiments that provide evidence of its truth.

5. A bar magnet is placed a short distance from a bar of soft iron. Their two axes are in the same straight line. Draw the magnetic field around the magnet and the bar. What is effect of this field upon the bar of soft iron?

6. What is meant by the terms (a) magnetic pole, and (b) magnetic field? State the inverse square law of magnetism.

Sketch the lines of force round (c) an isolated bar magnet, and (d) two north magnetic poles lying about 5 cm apart. (Ignore the effect of the Earth's magnetic field.)

7. Sketch the lines of force (a) around a bar magnet, and (b) between the adjacent north poles of two bar magnets.

8. You are given a nail and any other apparatus you may require. Describe experiments you would perform to discover if the nail is magnetized, and if it is, which end is the north pole.

How would you magnetize the nail so that the north pole is at the pointed end?

9. Describe how you would use a bar magnet to magnetize a steel knitting-needle so that its north pole is at a particular end. What is meant by saying that, at a certain place, the magnetic declination is 8° W and that the angle of dip is 65°?

10. What is meant by the angle of dip? What value would it have at the north magnetic pole?

11. What is meant by magnetic variation and how is it changing at the present time in this country? Draw a diagram to illustrate your answer.

12. Why, during the making of a steel ship, does it become magnetized?

13. Give a brief account of the earth's magnetism. Given a bar magnet how would you determine the magnetic meridan where you are?

26. Electricity

This subject is one of the most useful and exciting studies in the whole of physics. Although we do not know what electricity is, we know how to use it, we know what it can do, and we know many ways of producing it.

Electric charges. Let us start by finding out all we can about an electric charge formed by rubbing two things together. Take a strip of polythene about one inch wide and one foot long and pull it through a folded piece of woollen

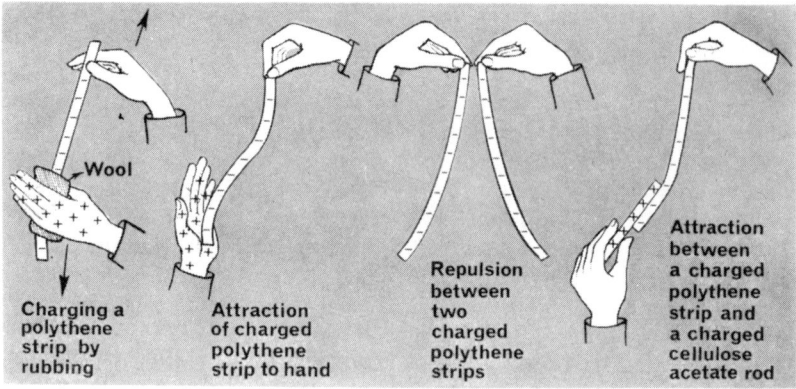

Charging a polythene strip by rubbing | Attraction of charged polythene strip to hand | Repulsion between two charged polythene strips | Attraction between a charged polythene strip and a charged cellulose acetate rod

Wool

Figure 26.1. The attraction and repulsion of electrically charged polythene strips.

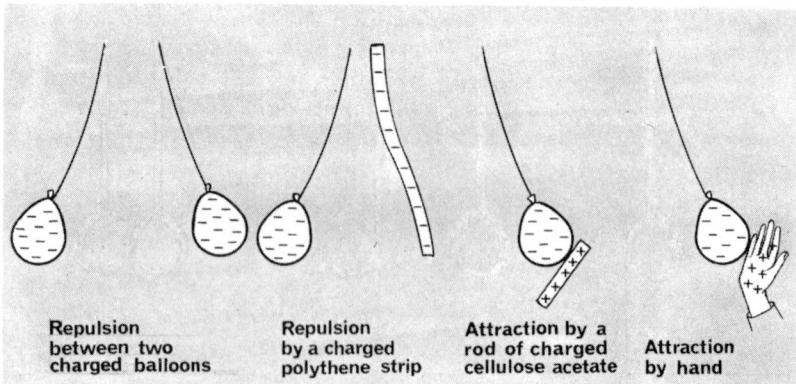

Repulsion between two charged balloons | Repulsion by a charged polythene strip | Attraction by a rod of charged cellulose acetate | Attraction by hand

Figure 26.2. How a negatively charged balloon behaves when other bodies are brought near to it.

material fairly rapidly. You will find that it is immediately attracted to the woollen material and to your hand. Two such strips of charged polythene hung closely together will repel one another very strongly. Take a rod of cellulose acetate and rub it vigorously with some silk. You will observe that it will attract dust and pick up small pieces of paper and other light objects. When you bring it close to the charged polythene strip there will be an immediate and very strong attraction between the two.

On the diagrams in the figures that follow, you will find a number of + and − signs marked on the charged materials. They represent the electric charges and it will become clear later why we use these particular signs and not any others. It will also be explained why some signs are drawn close together and others are separated.

Suspend two spherical rubber balloons each on a long length of silk thread. Rub them both with a piece of silk or with your dry hands. Observe the movements of the balloons when they are held close to one another. Observe also the movement of one of the suspended balloons as you bring towards it your own hand, the charged polythene strip, or the charged cellulose acetate rod used in the last experiment.

One important observation is that no charged body can be found that will be repelled by both the charged polythene strip and the cellulose acetate rod.

From these experiments we must conclude that electric charges are produced by rubbing because we saw and heard small electric sparks, that the same kind of charges are always produced when the same materials are rubbed, and that there is no third kind of charge. Two, and only two, kinds of charge exist and these are called positive and negative. We observe also that bodies with the same kind of charge (like charges) always repel one another and that bodies with different kinds of charges (unlike charges) attract one another.

An electroscope is a device that takes advantage of the repulsion between two similarly charged bodies to form a sensitive detector of electric charges. The bodies are two extremely thin leaves of gold mounted in a draught-proof glass case. When the leaves are charged they repel one another and move apart, and when the charges are removed the leaves close together again.

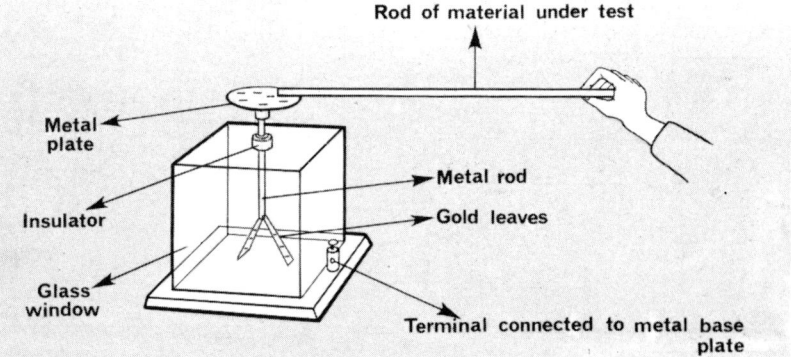

Figure 26.3. A charged electroscope being used to observe which materials are good and which are bad conductors of electric charges.

Charge an electroscope by gently stroking its metal plate with a charged polythene strip. In this way you will observe that some of the charges on the polythene strip are shared with the leaves of the electroscope. Test the *electrical conductivity* of a selected material by gently touching the metal plate of the charged electroscope with a rod of the material. Observe the rate of collapse of the leaves of the electroscope. *The good conductors of electricity* will allow the charges to escape quickly, causing a quick collapse of the leaves, whereas the bad conductors will allow them to escape slowly, and the insulators will not allow the charges to escape at all. Most metals are good conductors and most non-metals are insulators. Copper and aluminium are very good conductors of electricity. Sulphur, mica, polythene, and poly-vinyl chloride (P.V.C.) are good insulators. Most insulators, when their surfaces become damp, change into poor conductors and thus experiments involving small electric charges must be performed in dry conditions.

The study of conductors and insulators tells us why the charges on the polythene and cellulose acetate, both good insulators, shown in Figure 26.1 have been drawn separated from one another, whereas the charges on the hand, a good conductor, are drawn near together. In fact, the charges on a conductor can be made to move about as they are attracted or repelled by other nearby charges, whereas charges on insulators cannot move.

Electrostatic induction. This is the process of *inducing* two separate charges on a single insulated conductor. The process can be compared with magnetic induction. See Figure 25.7(*a*). The charges are produced where there were no charges before the operation. The presence of an inducing charge near an insulated conductor is sufficient to induce two equal and opposite charges on the insulated conductor. These two charges recombine and leave the conductor uncharged as soon as the inducing charge is removed. However, if the insulated conductor can be separated into two sections in the presence of the inducing charge the two separate charges can be isolated from one another. See Figure

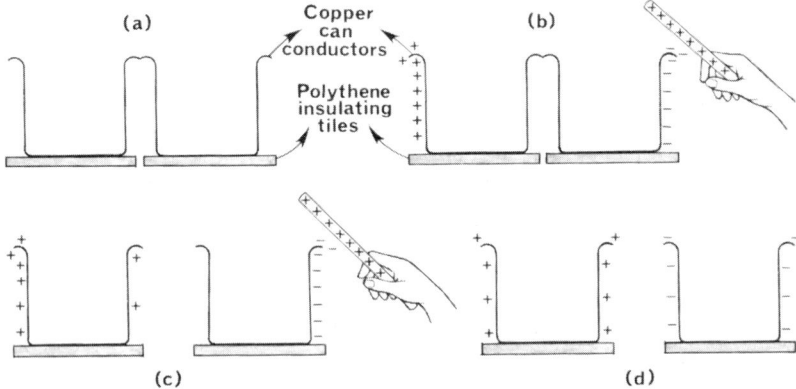

Figure 26.4. Stages in the charging of two conductors by induction. (*a*) Two uncharged and insulated metal conductors touch one another. (*b*) An inducing charge is placed close to one conductor. (*c*) The conductors are separated. (*d*) The inducing charge is removed.

26.4. If each of these two charges on their respective conductors are moved separately close to a charged electroscope, the leaves will diverge further in one case and collapse in the other indicating that they have different charges.

If, however, the two conductors are made to touch one another, the two charges neutralize one another and neither conductor will then disturb the leaves of the electroscope. The induced charges are therefore equal in quantity and opposite in effect to one another. As we have never found any evidence to suggest that there is any other kind of charge we therefore consider it sensible to name these two charges, the only kinds of charges, *positive* and *negative*. It is the usual custom to indicate the positive and negative charges by the mathematical symbols + and − as you will see has been done throughout this chapter.

The electron theory of matter. At the present time the accepted theory is that all matter consists of extremely small particles called molecules that cannot be divided further without losing their characteristic properties. All molecules contain atoms; for example, a water molecule consists of two atoms of hydrogen and one of oxygen. An atom contains a nucleus of *protons* and *neutrons*, and around this nucleus revolve *electrons* in the same way that planets revolve round the sun in the solar system. According to this theory the protons have a positive charge of electricity, the neutrons have no charge, and the *electrons have a negative charge*. Because each atom is neutral, it is thought that the number of protons in the nucleus equals the number of electrons orbiting around the nucleus in well-defined shells of different radii. *Any atom that loses one or more electrons becomes positively charged and an atom that gains one or more electrons becomes negatively charged.*

This electron theory enables scientists to explain many electrical observations and chemical reactions. Atoms can gain or lose electrons in many ways. As we have observed, by rubbing two materials together, we can make electrons move from atoms of one material to atoms of another. But it always requires some source of energy to cause such a flow of electrons. When we say that a *flow of electrons* takes place along a conductor in one direction we mean that the electrons themselves actually flow that way but, by convention, we have agreed that when we talk of an *electric current* we mean one of positive electricity flowing in the opposite direction. It is perhaps unfortunate that when it was first decided in which direction an electric current flowed the early scientists happened to chose the *direction opposite to the flow of electrons*.

We must now recall our first experiments concerning the rubbing of polythene strips and cellulose rods. It is supposed that electrons are rubbed off the atoms of the cellulose acetate, so this becomes positively charged because it has then too few electrons, and also that electrons are rubbed off the wool on to the polythene strip and thus the polythene strip becomes negatively charged having an excess of electrons.

Can the electron theory explain why copper is a good conductor of electricity? It is thought that the copper atom has its three innermost shells completely filled by orbiting electrons and that one extra electron orbits alone in the

N

fourth outermost shell. The theory suggests that this lonely electron is weakly bound to the atom and can easily be transferred to the next atom and the next and so on. In this way copper acts as a good conductor for the passage of an electron stream. It therefore easily permits the passage of an electric current.

What happens if a thick copper wire suddenly becomes constricted for a certain length? The free electrons passing easily along the thick wire from atom to atom then have to squeeze their way along a narrow passage and this causes great turmoil. This length of wire becomes hot, and it could even become red hot.

Distribution of charges on a conductor. How are the charges spread over one of the cans shown in Figure 26.4 after it has been charged and moved completely away from other charges? We can find this out by touching the can at different places by means of a very small insulated copper disc called a proof plane and taking off specimens of the charge. The size of the charge it picks up depends upon the density of the charge at the particular position touched on the surface of the can. See Figure 26.6. The size of the charge carried by the proof plane can be estimated by the movement of the leaves of an electroscope when the proof plane is brought up to touch the inside of a small can mounted on the plate of the electroscope in the way shown in Figure 26.7.

There are two observations of interest that can be made in this experiment. One is that there is no charge inside the conducting metal can. Thus if we wish to transfer the whole of a charge from one conductor to another all we have to do is to put the charged conductor *inside* and cause it to touch the uncharged one, and then remove it. All the charge is then transferred to the outside of the receiving conductor.

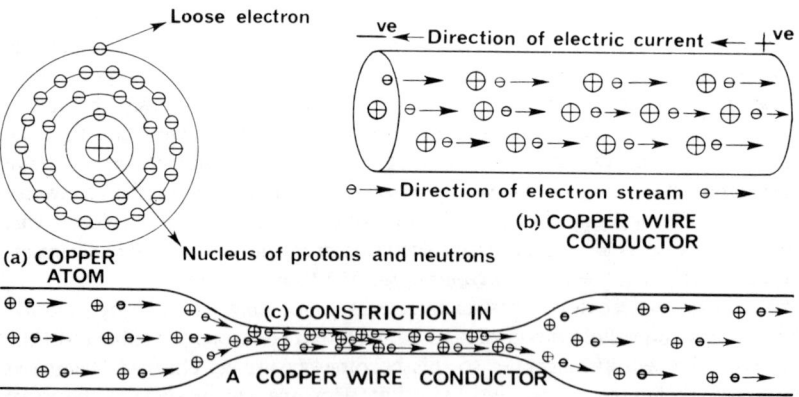

Figure 26.5. The conduction of an electric current along a copper wire. (*a*) A copper atom showing the position of the weakly bound electron, (*b*) shows how the free electrons move from atom to atom, and (*c*) shows how electrons have to squeeze their way through a narrow constriction in a conductor. The large circles in (*b*) and (*c*) with the + signs are the atoms that have lost the loose electron, and the small circles with the − signs are the electrons themselves. The atoms and electrons are, of course, more numerous than these diagrams indicate.

The other observation is the large concentration of the charges near the lip of the can. If there had been a point attached to the can, e.g. a pin soldered on, the number of charges on the point would have been greater still. In fact, the charges on a conducting point are so concentrated that they give their charges to the molecules of air near the point and these then repel one another so strongly that they form a vigorous air movement. This can be felt by the hand if a conducting point is connected to some large and continuous source of charges such as a Wimshurst or van der Graaff machine. The action observed at points explains why lightning conductors are so effective. The charges escaping by the air currents from the points above a building neutralize the charges on the cloud above.

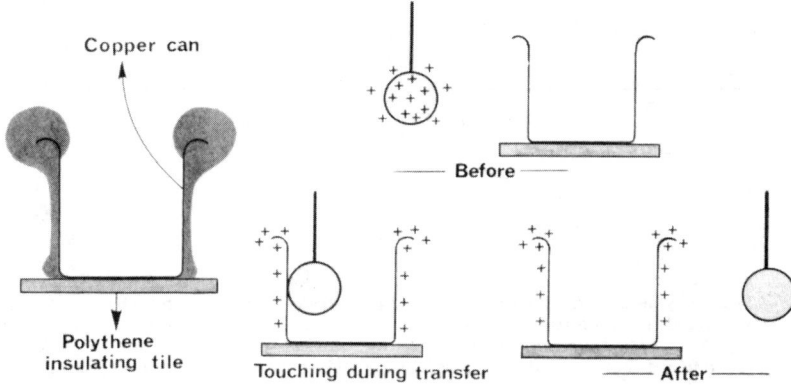

Figure 26.6. The shaded area represents the density of the electric charges at different places outside and inside the charged can.

Figure 26.7. This diagram shows how to transfer the whole charge of one conductor (a charged metal ball) to another (an uncharged metal can).

Figure 26.8. How lightning conductors help to protect a building from being struck by lightning. The conductor should be straight, thick, and connected to a big plate buried in the earth.

Some uses of an electroscope. If you have a charged insulated rod, such as an acetate or a polythene rod, you will find it is easier to charge an electroscope from it by the process of induction, as shown in Figure 26.9(*a*), than by stroking the charges off the rod. If you have a charged conducting body you can charge an electroscope by touching the two together and thus sharing the charges, see Figure 26.9(*b*), or better still by inserting the body into a deep metal can placed on top of the electroscope and transferring the whole charge of the body to the can and electroscope, see Figure 26.7.

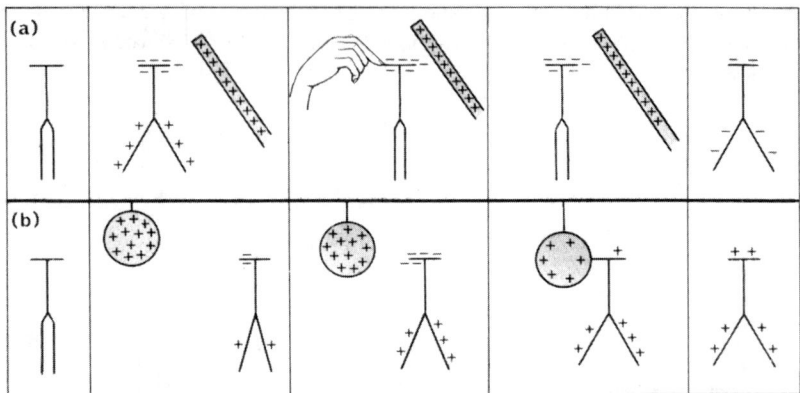

Figure 26.9. Stages in the charging of an electroscope. (*a*) Negatively by the use of a positively charged insulated rod. (*b*) Positively by touching it with a positively charged conducting body. The number of charges shown in these diagrams is only intended to indicate one possible distribution and to illustrate the general principles of the process.

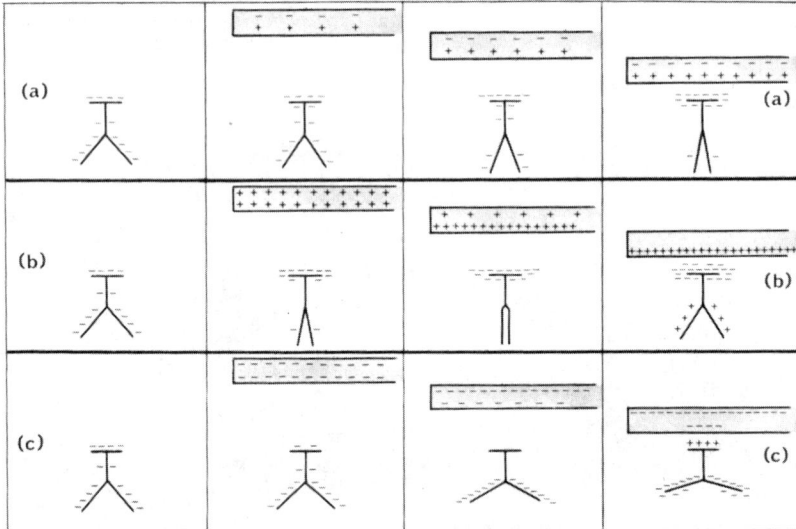

Figure 26.10. How the leaves of a negatively charged electroscope move as its plate is presented with (*a*) an uncharged body, (*b*) a positively charged body, and (*c*) a negatively charged body. How would a positively charged electroscope behave under these same three conditions?

A charged electroscope can be used to determine if a body moved steadily towards it is uncharged, charged positively, or charged negatively. All one has to do is to know whether the electroscope is positively or negatively charged and then to observe the movements of the leaves of the electroscope. The reasons for the movements of the leaves will be clear after you have examined Figure 26.10.

Sources of electricity. So far we have studied electric charges produced by friction. The Greeks knew of these frictional charges. They rubbed amber and found that it would then attract small light objects. Our word 'electric' is derived from the Greek word 'elektron' meaning amber. There are other sources of electric charges and we shall now consider them.

Electricity from chemical energy. The most satisfactory primary cell in general use is that formed by zinc and carbon plates immersed in ammonium chloride (sal-ammoniac). Eventually this cell becomes exhausted when all the zinc is turned to zinc chloride by the action of the ammonium chloride and the zinc case falls to pieces. When the cell is connected in a *closed electric circuit* so that a current flows the chemical action consumes the zinc and at the same time covers the carbon rod with bubbles of hydrogen. These hydrogen bubbles are removed slowly by the manganese dioxide and then the cell is ready to function again. Thus these cells are only suitable for intermittent operation, their output is limited, and they are a costly way of producing electrical energy.

The flow of electrons round a conducting circuit occurs from the negative terminal (usually coloured black) to the positive terminal (red) of the cell. The conventional 'electric current', passing in a direction opposite to the flow of electrons, therefore leaves the positive terminal of the cell and after travelling all the way round the electric circuit returns to the negative terminal.

When primary cells are 'run down' they can only work again if the chemicals are replaced but *secondary* cells can rebuild their chemicals if an electric current is passed through them in the opposite direction to the original

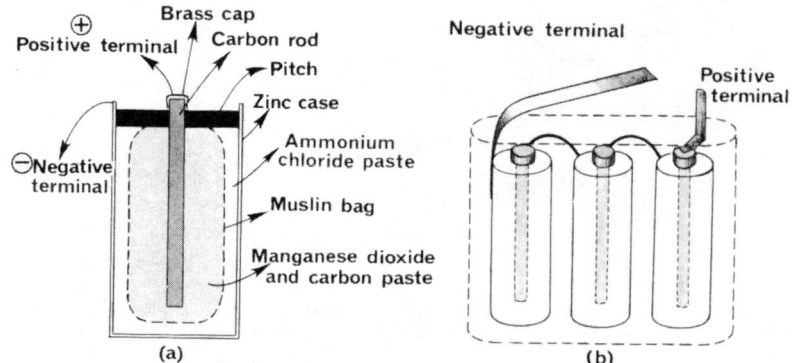

Figure 26.11. The Leclanché cell. (*a*) The 'dry' cell. Around the carbon rod is a bag containing manganese dioxide and carbon; these prevent the formation of the hydrogen bubbles that would otherwise form on the rod owing to the action taking place when the cell is in use. (*b*) A battery of three dry cells.

discharge current, by a process known as re-charging. An accumulator is a secondary cell. It stores energy by converting electrical into chemical energy and then during the process of discharge changes it back to electrical energy. The type used commonly in motor-cars consists of lead plates immersed in sulphuric acid. (See Figure 26.12.)

Electricity from mechanical energy. Another method of producing electricity is the *generator* that transforms the mechanical energy used to rotate its axle into electrical energy. Some generators can produce an electric current having different properties from that produced by frictional or chemical means.

An electric current flows from a positively charged body to a negatively charged body, and from the positive terminal to the negative terminal of a cell through the outside circuit. Some generators also produce a similar electric current, i.e. one that flows in one direction only. This electric current is known as a *direct current* (d.c.). The stream of electrons in each case is again in the opposite direction to the electric current.

There is another type of generator that produces an electric current that surges to and fro in its circuit. This current has different properties and can be used for different purposes. It is called an *alternating current* (a.c.). In this alternating current the stream of electrons repeatedly reverses its direction.

Generators produce electricity by the movement of a magnet relative to a coil of conducting wire and the amount of the electricity produced depends on the rate of rotation, the size of the machine, the strength of the magnet, and the number of turns in the coils of wire.

We have now studied three ways of producing electric currents. We have learned that an electric current is a movement of electrons. It is not easy at first to realize that the electric charges produced by friction and the electric charges that flow in a wire connected to a battery are really both concerned with electrons. They are exactly the same but they act in different ways and because

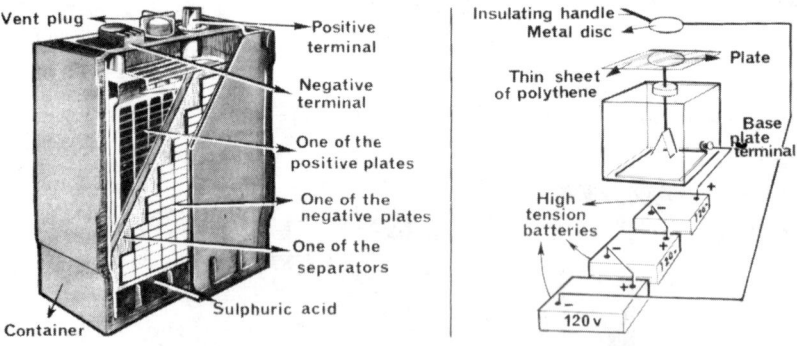

Figure 26.12. An accumulator. The positive and negative plates have a large surface area and often consist of many layers sandwiched together and separated by thin perforated insulating sheets.

Figure 26.13. How primary cells can cause the leaves of an electroscope to diverge. Each battery contains many cells. The cells provide negative charges on the metal disc and positive charges on the base plate. These charges induce positive charges on the plate and negative charges on the leaves of the electroscope.

of this they appear different. The essential difference is that frictional electric charges are produced at very high *potentials* and can be seen jumping across gaps as sparks whereas those produced by chemical and mechanical means are usually at much lower potentials. We shall learn what is meant by potential (voltage) later. However, high-potential sparks can be produced from cells by means of special equipment. The battery of a motor-car provides the energy for the sparks formed at the plugs of the engine. The equipment which does this is called an ignition coil. We can demonstrate that the electric charges produced by a large number of cells can cause the leaves of an electroscope to diverge in the same way that frictional electric charges can. (See Figure 26.13.)

The Electron

Thus frictional electric charges (static electricity) and electric charges produced by chemical and mechanical means (current electricity) are the same form of energy. Electricity is a form of energy because electricity is produced only when other forms of energy are consumed.

Still a fundamental question is unanswered. What is an electron? We know it is a form of energy, but are there any other properties we can find out about it?

Negative electric charges (electrons) are given off when a filament of tungsten is heated strongly. It is as if they were 'boiled' off the metal filament. Most of these electrons can be collected on a metal plate if (*a*) the metal plate is connected to the positive terminal and the filament to the negative terminal of a battery, and (*b*) the metal filament and plate are enclosed in a vacuum tube.

We can use similar streams of electrons in other specially designed 'diode' tubes to show that electrons have certain properties. Look at the diodes illustrated in Figure 26.15. In each case the filament is inside the tube on the left and it is heated by passing an electric current through the external wires

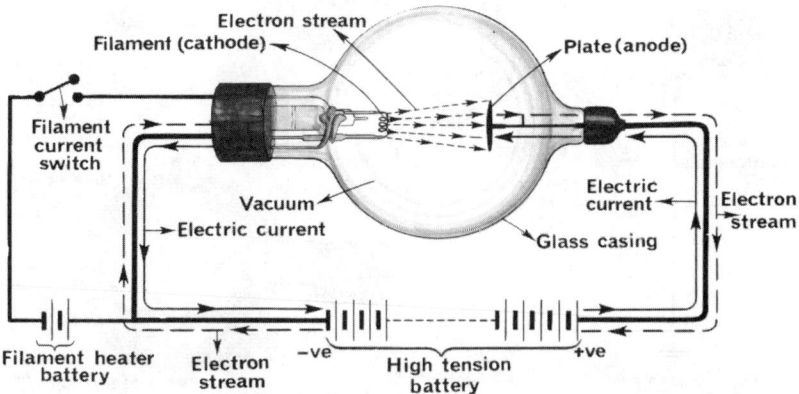

Figure 26.14. **A** diode. There are two items in this vacuum tube, the filament and the plate. Electrons from the heated filament are attracted to the plate and continue round the circuit following the dotted arrows. The continuous arrows represent the passage of the electric current from the positive terminal to the negative terminal of the high tension battery.

shown. Electrons are 'boiled' off this filament and are then attracted to the plate. The plate is often called the *anode* because it is connected to the positive of a high tension battery. The negative of the same high tension battery is connected to the filament which is why it is often called the *cathode*. Hence the electrons emitted by the 'hot' cathode travel to the anode, then pass through the high tension battery, and complete the circuit by returning to the cathode as shown also in Figure 26.14. It is, of course, the potential or voltage applied by means of the high tension battery that causes the electrons to flow round the circuit.

In tube (*a*) some of the electrons attracted to the anode pass through the hole in the anode and travel on as a straight beam of electrons. This beam follows the dotted path and produces a bright spot on the fluorescent patch on the wall of the tube. The beam of electrons is bent along a curved path when it passes through a magnetic field. The stronger the magnetic field the more the beam curves. The continuous path in the drawing shows the bending of the electron beam when a long bar magnet is brought close enough to the tube to cause the beam to enter a metal collecting cup. When this happens the electrons collected pass along the wire from the cup to the electroscope and a movement of the leaves of the electroscope is observed. The direction of this movement of the leaves of the negatively charged electroscope indicates that negative electric charges accompany the arrival of the electrons. Millions and millions of these electrons travel along a television tube and although we cannot see them we can see the tiny speck of light they produce when they hit the chemicals on the screen at the end of the tube which we look at.

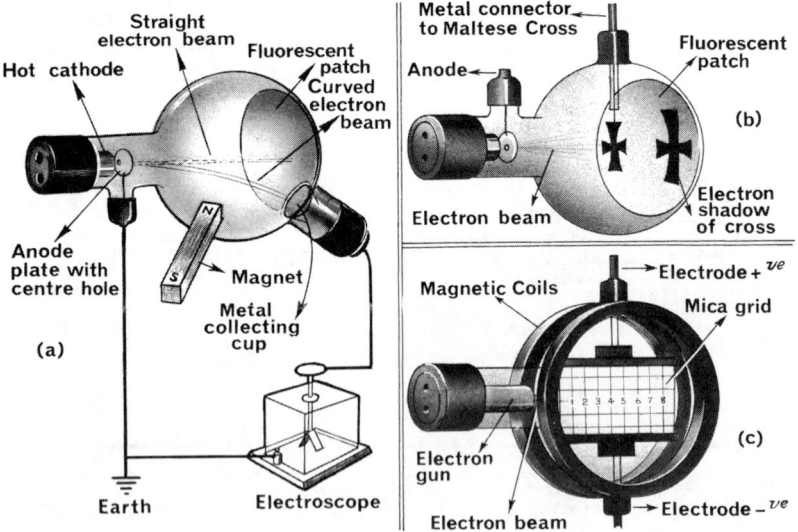

Figure 26.15. Three diode tubes designed to produce narrow beams of elect·ons. The cathode in each case is provided by a hot glowing filament connected in an electric circuit by the two external wires. Each tube can be used to demonstrate some of the properties of a beam of electrons.

In tube (b) the electrons form a shadow of the 'Maltese cross' on the fluorescent patch on the wall of the tube. This shadow is in exactly the same place as the linear shadow of the cross formed by the light of the glowing filament before the anode is connected to the positive terminal of the battery. Because light travels in straight lines it is reasonable, therefore, to assume that electrons also travel in straight lines. So we have two shadows, one a light shadow and the other an electron shadow. Moreover one can bend the path of the electrons by bringing up a magnet close to the tube as one can in tube (a) and obtain a distorted electron shadow. If however the 'Maltese cross' is disconnected from the anode entirely the first few electrons settle on the cross and cannot escape. They then repel further electrons as they arrive near the cross and one sees a shadow which is bulging out on all its sides. This confirms that electrons are charges of electricity which repel one another—we have learnt this earlier that like charges repel one another.

Tube (c) enables us to observe and calculate the amount of deflection caused by magnetic and electric fields on an electron. The straight line beam of electrons passes between two metal plates, one charged negatively, and the other positively. The electric field so produced deflects the beam. The beam just scrapes along a thin mica sheet on which the grid of a graph is printed.

The path of the beam so deflected bends over the grid, and from its trace one is able to observe the amount of deflection produced by a certain potential difference across the two charged plates.

A similar deflection is produced when a uniform magnetic field is created by the two coils outside the tube. The design of these coils is such that the magnetic field is uniform. Again the amount of deflection can be observed by the trace on the mica grid.

The amount of bending in both cases is caused by the strength of the electric or magnetic field and can be upwards or downwards according to the direction of the fields.

Thus whatever electrons may be they have negative charges and they travel in straight lines. They can also be deflected in the form of an arc when they travel through a magnetic field in the same way that a thin light aluminium strip is bent when an electric current passes through it in a magnetic field. Look at Figure 28.1. This is further evidence that a beam of moving electrons has the same properties as an electric current.

The modern theory of the structure of the atom thus leads us to understand that electricity is a form of energy, that the absence or the presence of some things we call electrons cause bodies to possess positive or negative charges of electricity, and that an electric current flows when these same electrons move. But even all this does not tell us precisely what an electric current is because we do not know the real nature of an electron.

Questions on chapter 26

1. Describe an experiment to show that electric charges can be produced by friction.

2. Why do we think that there are two and only two different kinds of electrical charge?

3. Describe an experiment by which you could show that like charges repel and unlike charges attract one another.

4. How does the force of attraction or repulsion between two electrical charges vary with the size of the charges and with the distance between the charges?

5. Are charges created when two bodies are rubbed together? Explain what takes place, including the condition of each body before and after being charged.

6. Draw a diagram of a gold leaf electroscope and label the parts which are insulators or conductors. How would you use the electroscope to find out if a given piece of material is an insulator or a conductor?

How would you charge the electroscope with positive electricity using a negatively charged rod?

7. What is the difference between a conductor and an insulator? Give one example of each.

8. What is meant by electrostatic induction? Show by a series of diagrams how you would charge an insulated metal sphere positively by induction.

9. How would you use an electroscope to show that a body possesses a positive charge?

10. Explain the action of a lightning conductor. Draw a diagram to show what happens when a charged cloud passes over it.

11. Why are the 'points' on a sparking plug sharpened?

12. How are the following explained by means of the electron theory: (a) a neutral body, (b) a negatively charged body, and (c) a positively charged body?

27. Electrical measurements

In some ways an electric current can be compared with a water current and with a gas current. All three pass along conducting 'pipes' and are forced along these pipes by some form of pressure against obstructions that control their passage. It is possible to measure the quantity of water flowing along a pipe by so many bucketfuls every second of time. Likewise it is possible to measure the *quantity* (Q) of electricity flowing along a wire by the number of *coulombs* that pass every second. The quantity, or charge, of one coulomb of electricity contains approximately 6.3×10^{18} electrons or nearly two thirds of ten million, million, million electrons. This gives us an idea of the extremely minute charge of a single electron in an atom. If one coulomb of electricity flows along a wire every second then the rate of flow of the electric *current* (I) is one *ampere*. The ammeter measures the rate of flow of electrons or electric current.

CHARGE		=	CURRENT		×	TIME
A charge			a current			a time
or of	passes when		of	flows for		of
1 coulomb			**1 ampere**			**1 second**

The difference in pressure required to force water along a pipe corresponds to the *potential difference* (V) required to force a charge of electricity along a wire. The unit of potential difference, often contracted to *potential*, is a *volt* and thus we speak of *voltage* as the potential difference that keeps a current

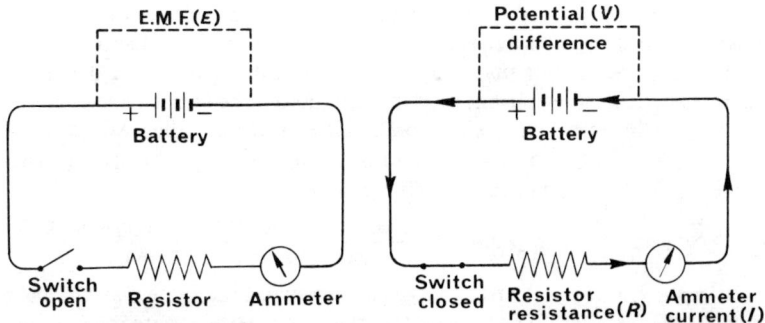

Figure 27.1. A simple electrical circuit. The battery of cells provides the electromotive force (E) ready to drive a current around the circuit. The switch completes the circuit and a current (I) starts to flow from the positive terminal of the battery through the switch, resistor, and ammeter to the negative terminal of the battery, and then through the battery itself to the positive terminal. When the current flows the battery maintains a potential difference (V) across its terminals.

flowing in a circuit. We speak, too, of the *electromotive force* (*E*), also measured in volts, as being the potential of the supply that is ready to force electricity along a circuit. The instrument that measures potential is a voltmeter.

If the potential difference forces a charge of electricity along a circuit then electrical energy is consumed. The energy does not disappear. It is transformed into some other form of energy—heat energy or light energy or perhaps mechanical energy. The unit of electrical energy is the *joule* and the following equation applies:

ENERGY	=	CHARGE	×	POTENTIAL
An energy		a charge		a potential
or of is consumed when		of passes across		difference of
1 joule		**1 coulomb**		**1 volt**

To keep a current of electricity flowing continuously along a circuit, electrical power is needed. The power depends both on the potential difference and the current. The unit of electrical power is the *watt*. This can be expressed by the following equation:

POWER	=	CURRENT	×	POTENTIAL
A power		a current		a potential
or of is needed when		of flows across		difference of
1 watt		**1 ampere**		**1 volt**

From these units we see that 1 joule of energy is that provided by a power of 1 watt for 1 second. These are small units when considered in terms of the electrical energy needed for domestic heating and lighting. Hence we buy electrical energy in much larger units called kilowatt-hours (kWh). This is the energy supplied by 1 000 watts of power for 1 hour, or $60 \times 60 \times 1\,000$ volt-ampere-seconds (joules).

In the case of water, obstructions in the pipe control the rate of flow of water from the end of the pipe if the pressure of water between the two ends remains constant. In the electrical case the *resistance* (*R*) of the conductors in a circuit determines the current that a certain voltage can drive round the circuit. In fact, the greater the resistance of the conductors the smaller the current. The unit of resistance is the *ohm*, abbreviated as Ω; in other words a *resistor* of resistance 1 ohm allows a current of 1 ampere to flow through it when the potential difference between its ends is 1 volt.

$$\text{CURRENT (in amperes)} = \frac{\text{POTENTIAL DIFFERENCE (in volts)}}{\text{RESISTANCE (in ohms)}}$$

A *capacitor* stores electric charges in somewhat the same way that a sealed container stores gas. The greater the gas pressure the greater the quantity of gas that can be pressed into it. So the greater the potential applied across the plates of the capacitor the greater the charge or quantity of electricity that can be stored by it. If a potential difference of 1 volt applied to a capacitor stores in it 1 coulomb of electricity it is said to have a *capacitance* of 1 *farad*.

CHARGE	=	CAPACITANCE	×	POTENTIAL

	A charge		a capacitance		a potential
or	of	is stored by a capacitor having	of	when across its plates there is	difference of
	1 coulomb		**1 farad**		**1 volt**

A capacitor of capacitance 1 farad is very large indeed; those in use in radio work are measured in 10^{-6} farads and 10^{-12} farads and these units of capacitance are called *microfarads* (μF) and picofarads (pF).

Electrical terms	Units	Symbols
Charge	coulomb, C	Q
Current	ampere, A	I
Electromotive force		E
Potential difference	volt, V	V
Potential		V
Voltage		V
Energy	joule, J	W
	kilowatt-hour, kWh	W
Power	watt, W	P
Resistance	ohm, Ω	R
Capacitance	farad, F	C

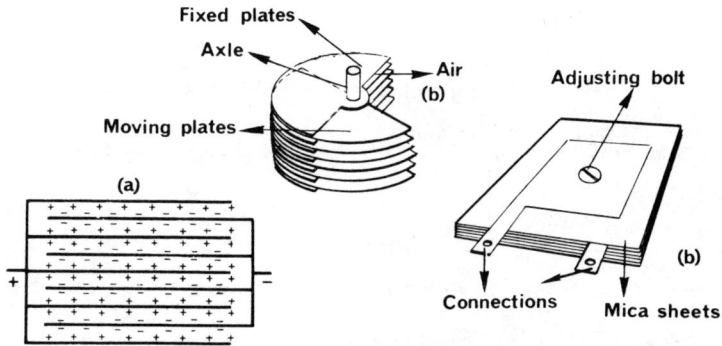

Figure 27.2. Some types of capacitor. The electric charges are held on the surfaces of the plates. (*a*) The fixed capacitor has many plates separated by thin sheets of insulating material and all of these are packed closely together. (*b*) The variable capacitor. The plates of one of these move in and out of one another in air, and the plates of the other, separated by thin mica sheets, are squeezed together by a nut and bolt.

The fundamental relationship between the current, potential difference, and the resistance is known as *Ohm's Law*.

$$\text{CURRENT (amps)} = \frac{\text{POTENTIAL (volts)}}{\text{RESISTANCE (ohms)}}$$

This relationship applies to single resistors in a circuit or to parts of a circuit. It can apply also to a whole circuit, in which case the e.m.f. of the cells is the 'potential' which forces the current through the resistance of all the resistors in the whole circuit.

Series circuit. Suppose a generator is connected to three lamps in series. All the current that flows round the circuit has to pass through one lamp, then through the next lamp, and also through the third lamp before it returns through the generator to its starting terminal. This is what is meant by a series circuit. The total resistance offered by the three lamps in series (each lamp is a resistor) is the sum of the resistances of the three lamps taken separately. So that this equation applies:

$$R = R_1 + R_2 + R_3$$

Let us suppose that the generator supplies a constant e.m.f. of 90 volts and that the three lamp resistors have resistances of 9, 6, and 3 ohms respectively. The resistance of the whole circuit can be calculated thus:

$$R = R_1 + R_2 + R_3,$$
$$\therefore R = 9 + 6 + 3,$$
$$\therefore R = 18.$$

Therefore the resistance of the whole circuit is 18 ohms.

The current flowing through the whole circuit can be calculated using the value of the resistance of the whole circuit thus:

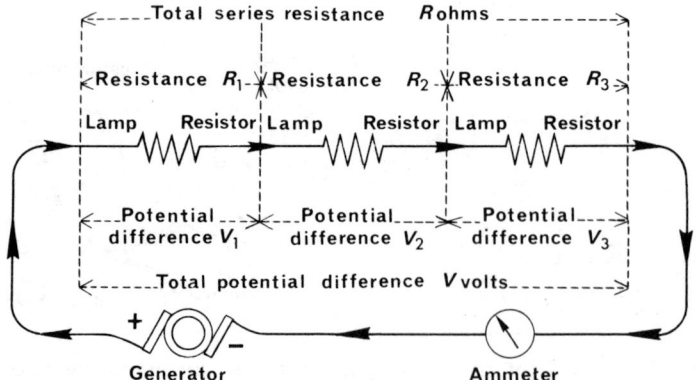

Figure 27.3. The wiring diagram of a series circuit including three lamps and a generator. The dotted lines show the potentials and the resistances between their respective connections.

$$I \text{ (amps)} = \frac{E \text{ (volts)}}{R \text{ (ohms)}},$$

$$\therefore I = \frac{90}{18},$$

$$\therefore I = 5.$$

Therefore the current flowing through the whole circuit is 5 amperes.

The potential differences, V_1, V_2, and V_3, across each of the lamp resistors can now be calculated using the value for the current flowing through each lamp resistor in the whole circuit thus:

$$I \text{ (amps)} = \frac{V \text{ (volts)}}{R \text{ (ohms)}},$$

or $V \text{ (volts)} = I \text{ (amps)} \times R \text{ (ohms)}$.

$\therefore V_1 = 5 \times 9$, and $V_2 = 5 \times 6$, and $V_3 = 5 \times 3$.

$\therefore \underline{V_1 = 45}$,   and $\underline{V_2 = 30}$,   and $\underline{V_3 = 15}$.

Therefore the potential differences across each of the lamp resistors are 45 volts, 30 volts, and 15 volts respectively.

These calculations have given us all the values of the currents, potentials, and resistances in the whole and the parts of this circuit. We observe that the potential difference in each section along a series circuit is always proportional to the resistance of that particular section.

If instead of having three lamps in series there had been any other number of lamps the calculations could be made in exactly the same way using the equation:

$$R = R_1 + R_2 + R_3 + R_4 + R_5 + \ldots$$

A series circuit is used for connecting a string of 'fairy lamps', for supplying a low voltage to each of the valves in a radio set, and for starting the electric motors of some locomotives until they have gained speed and are switched over to parallel circuits.

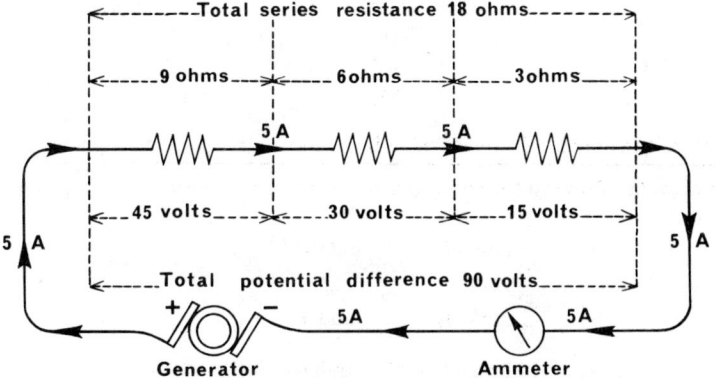

Figure 27.4. The complete information concerning the current, potential, and resistance of this series circuit.

Parallel circuit. Suppose a generator is connected to three lamps in parallel. The circuit is completed through three different branches, one through each lamp resistor, and the total current taken from the generator must therefore be the sum of the currents taken in each branch. In other words in the parallel circuit we can add the currents flowing through each branch whereas in the series circuit we added the resistances of each resistor.

Using Ohm's Law and the values shown in the diagram of Figure 27.5:

$$I = I_1 + I_2 + I_3,$$

$$\text{or } \frac{E}{R} = \frac{E}{R_1} + \frac{E}{R_2} + \frac{E}{R_3},$$

$$\text{or } \frac{1}{R} = \frac{1}{R_1} + \frac{1}{R_2} + \frac{1}{R_3}.$$

Let us suppose that the generator supplies a constant potential of 90 volts and that the three lamp resistors in parallel have resistances of 9, 6, and 3 ohms respectively. The resistance of the whole circuit can be calculated thus:

$$\frac{1}{R} = \frac{1}{R_1} + \frac{1}{R_2} + \frac{1}{R_3},$$

$$\therefore \frac{1}{R} = \frac{1}{9} + \frac{1}{6} + \frac{1}{3},$$

$$\therefore \frac{1}{R} = \frac{11}{18},$$

$$\therefore R = 1{\cdot}6.$$

Therefore the resistance of the whole circuit is 1·6 ohms.

The current flowing through the whole circuit can be calculated using the value of the resistance of the whole circuit thus:

$$I \text{ (amps)} = \frac{E \text{ (volts)}}{R \text{ (ohms)}},$$

$$\therefore I = \frac{90}{1{\cdot}6},$$

$$\therefore I = 55.$$

Therefore the current flowing through the whole circuit is 55 amperes.

The current flowing through each lamp resistor can be calculated thus:

$$I_1 = \frac{E}{R_1}, \text{ and } I_2 = \frac{E}{R_2}, \text{ and } I_3 = \frac{E}{R_3},$$

$$\therefore I_1 = \frac{90}{9}, \text{ and } I_2 = \frac{90}{6}, \text{ and } I_3 = \frac{90}{3},$$

$$\therefore I_1 = 10, \text{ and } I_2 = 15, \text{ and } I_3 = 30.$$

Therefore the currents flowing through each of the lamp resistors in parallel are 10 amperes, 15 amperes, and 30 amperes respectively.

These calculations have given us all the values of the currents, potentials, and resistances in the whole and the branches of these circuits. We observe that the sum of the currents in the branches, (10 + 15 + 30) amps, is the same as the current flowing through the whole circuit, 55 amps.

If instead of having three lamps in parallel there had been any other number of lamps the calculations could have been made in exactly the same way using the equation:

$$\frac{1}{R} = \frac{1}{R_1} + \frac{1}{R_2} + \frac{1}{R_3} + \frac{1}{R_4} + \frac{1}{R_5} \cdots$$

One of the great advantages of the parallel circuit is that any one resistor can be switched out of the circuit without interfering with the current flowing through the other resistors. Most electrical circuits in the house have parallel circuits for the lamps, fires, and other appliances. Main fuses, meters, and main switches are connected in series with the parallel circuits.

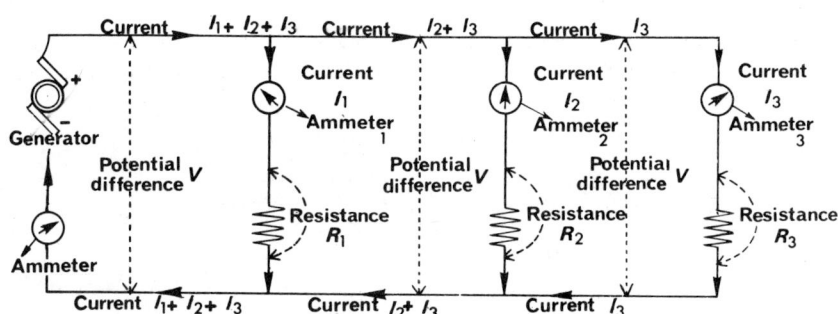

Figure 27.5. The wiring diagram of a parallel circuit including three lamps and a generator. The dotted lines show the potentials and the resistances between their respective connections.

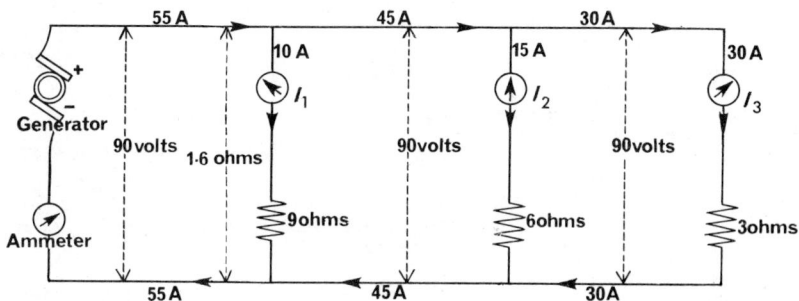

Figure 27.6. The complete information concerning the current in each section of the circuit, the potential across and the resistance of all the parallel circuits.

O

Factors controlling resistance. From all we have so far learned about resistors it is clear that if we had, for example, two equal resistors and put them in series the total resistance of the two together would be twice the resistance of one of them. If, however, we put them in parallel the total resistance would be one half the resistance of one alone. If we had not two equal resistors but, for example, six equal resistors the total resistance in series would be six times, and in parallel would be one sixth, the resistance of one single resistor. This can be expressed in mathematics by saying that in the case of a length of uniform wire:

1. the resistance is proportional to the length, or

$$R \propto l,$$

2. the resistance is inversely proportional to the cross-sectional area, or

$$R \propto \frac{1}{A},$$

and by combining these into one equation we can write it thus:

$$R = \frac{\rho l}{A}.$$

The constant ρ in this equation depends only on the material and the physical conditions of the resistor wire and it is called the *resistivity* of the material.

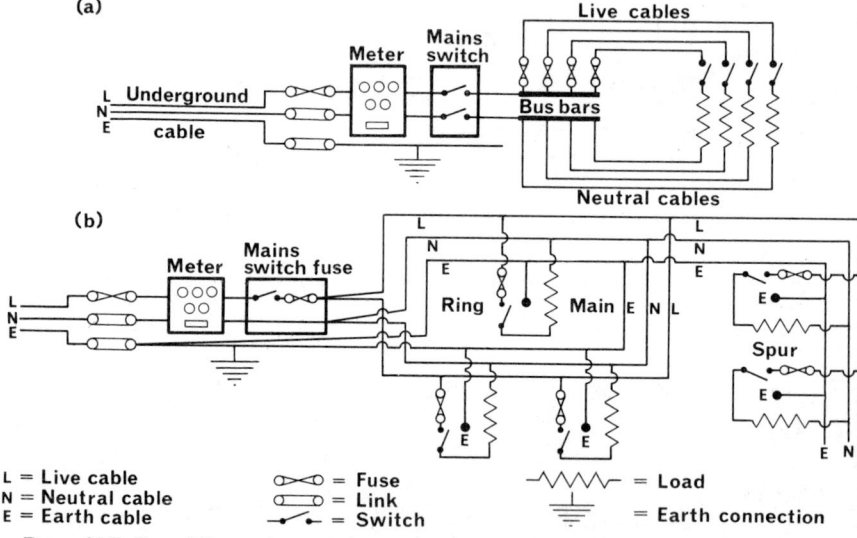

Figure 27.7. Two different domestic house circuits. Parallel circuits for the separate loads are used in both types. Switches and fuses are in the LIVE cables only.

(a) The 'distribution' fuse box type in which each parallel circuit is connected to bus bars. For the sake of clarity the EARTH cable has not been drawn around the parallel circuits.

(b) The 'ring' circuit type in which the individual fuse for each parallel circuit is included inside its plug. A spur, which is used for no more than 6 light socket outlets, is drawn. This type of circuit uses less cable than type (a).

In both types a faulty appliance 'blows' its particular fuse and thus does not cause the whole system to become 'dead'.

It can be defined as the resistance from one face to another of a cube of the material 1 cm long and of 1 sq.cm. cross-sectional area. (Figure 27.9.). It is measured in units of *ohm cm*.

Cells in series. When several cells are connected in series in a circuit in order to supply the electrical energy (in place of the electric generator used in our previous examples) the total electromotive force of all the cells is equal to the sum of the separate electromotive forces of each cell. A cell offers some resistance to an electric current flowing through it and this is known as its *internal resistance* (r). The internal resistance of a number of cells in series is equal to the sum of the internal resistances of each cell. Expressed in the form of equations these two statements, using the symbols shown in Figure 27.10 are:

$$E = E_1 + E_2 + E_3 + E_4 + E_5 + E_6 + \dots$$

and
$$r = r_1 + r_2 + r_3 + r_4 + r_5 + r_6 + \dots$$

One of the main reasons why cells are connected in series is that they can thus provide a greater electromotive force than that provided by a single cell. For some purposes it may be necessary to apply a large potential difference at the terminals of the equipment to be used. For instance, the lamps of many motor-cars will only light properly when the potential difference applied is 12 volts. One accumulator of 2 volts could not do this and thus a battery of six accumulators in series is used instead.

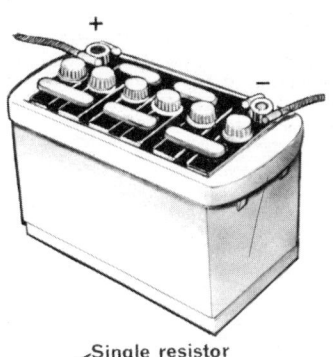

Figure 27.8. A motor-car battery of 12 volts consists of six separate accumulators each of 2 volts connected in series. No matter how large you make one accumulator cell it will still only have an e.m.f. of 2 volts.

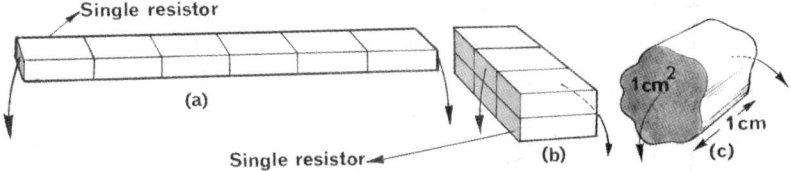

Figure 27.9. Resistors. Six equal resistors (*a*) in series, and (*b*) in parallel. The resistances between the faces marked with the arrows are six times and one-sixth of the resistance of one of the resistors. (*c*) A resistor of 1 cm long and 1 sq cm cross-sectional area.

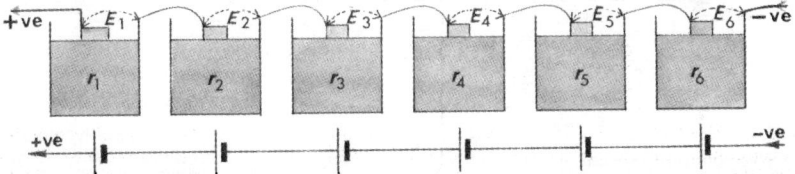

Figure 27.10. A number of cells connected in series. The electromotive force (E) and the internal resistance (r) of each cell is shown. Any number of cells can be connected in this manner. The graphical symbols as used in circuit and wiring diagrams are shown below.

Cells in parallel. When several cells are connected in parallel as shown in Figure 27.11 all the positive terminals are joined together, and all the negative terminals are joined together. We are creating in this way a single large cell whose plates are equal in area to the sum of the areas of the plates in all the individual cells. The total internal resistance of such an arrangement of cells will be less than the internal resistance of any one of the single cells. If four identical cells are connected in parallel the internal resistance is reduced to one-quarter of the internal resistance of a single cell, but the electromotive force is unaltered. The internal resistance can be calculated by the general equation:

$$\frac{1}{r} = \frac{1}{r_1} + \frac{1}{r_2} + \frac{1}{r_3} + \frac{1}{r_4} + \frac{1}{r_5} + \ldots$$

One of the main reasons why cells are connected in parallel is that the demand for a heavy current can be spread in this way among several cells. A heavy current taken from a single cell may damage the cell or shorten its life. Accumulators should *never* be connected in parallel. Can you explain why not? (The internal resistance of an accumulator may be as low as 0.01 ohm.)

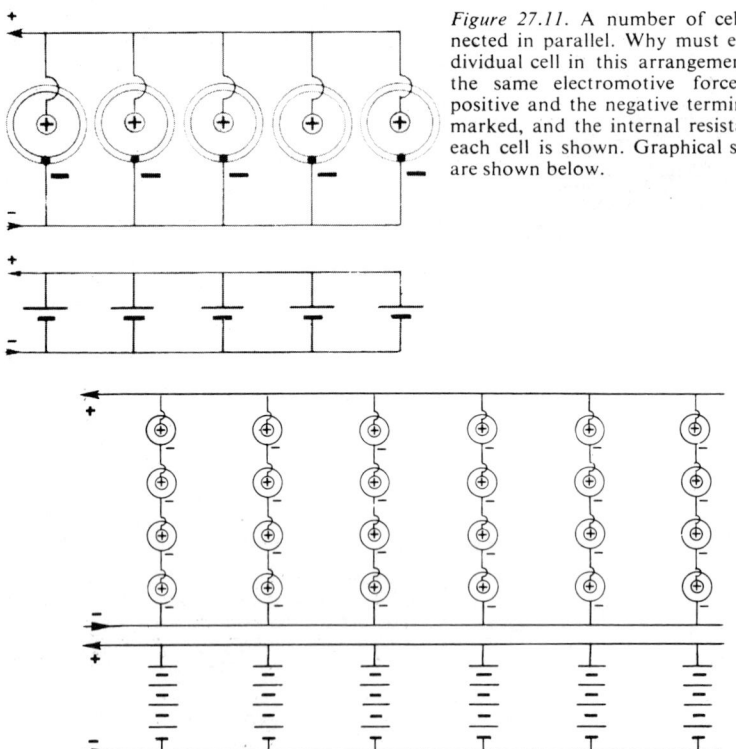

Figure 27.11. A number of cells connected in parallel. Why must each individual cell in this arrangement have the same electromotive force? The positive and the negative terminals are marked, and the internal resistance of each cell is shown. Graphical symbols are shown below.

Figure 27.12. A series-parallel arrangement of identical cells. There are six parallel circuits each containing four cells in series. Why is the potential difference four times and the internal resistance four-sixths that of the respective values for the single cell? Graphical symbols are shown below.

Arrangements of cells. A number of identical cells can be grouped together either in series or in parallel depending on whether we desire a high electromotive force or a low internal resistance. Banks of cells in series can also be grouped together in parallel as shown in Figure 27.12. A number of cells grouped together in any combination is called a *battery*. When it is desired to obtain the greatest power from a battery, the cells must be connected together so that the internal resistance of the battery so formed equals as nearly as possible the external resistance present in the external circuit formed by the equipment in use. If the external resistance is large, the cells should be joined in series so as to provide a high electromotive force. If the external resistance is low, the cells should be joined in parallel; this will keep the total internal and external resistance low and thereby enable a large current to flow.

Questions on chapter 27

1. What do the ammeter and the voltmeter measure and how are they connected in a circuit?

2. Describe how you would use a voltmeter and an ammeter to measure the electrical resistance of an immersion heater.

3. State Ohm's law. Describe how you would use this law to measure the resistance of a piece of wire; draw a diagram of the circuit you would use.

4. An electric light bulb is connected in series with a rheostat across a d.c. supply. Draw a circuit diagram of this arrangement including an ammeter to measure the current through the bulb and a voltmeter to measure the potential difference across it.

5. Resistors of 2 ohms and 8 ohms are connected (a) in series, and (b) in parallel. Calculate the effective resistance in each case.

6. Three 2 volt cells are connected (a) in series, and (b) in parallel. What would be the voltage of the battery formed in each case? What is the main advantage of an accumulator as compared with a primary cell?

7. What is meant by electromotive force and internal resistance?
A cell has an internal resistance of 1 ohm, and is connected to a resistor of 2 ohms. The current in the circuit is found to be 0.5 ampere. What is the e.m.f. of the cell?

8. Why is it desirable that the resistance of an ammeter be as low as possible and the resistance of a voltmeter be as high as possible?

9. Why are the normal lamps and heaters used in a house electrical circuit connected in parallel and never in series?

10. Some lamps especially designed for decorating Christmas trees are marked 20 V, 0.3 A. How many should be connected in series on to a 240 V mains supply? What power do they require when lighted?

11. Are the headlamps on a motor-car connected in series or in parallel? What evidence could an experienced motorist supply to show that your answer is correct?

28. The magnetic effects of an electrical current

If we observe carefully the movements that take place when the two simple experiments shown in Figure 28.1 are performed, we shall learn that (1) forces are created between a magnet and a wire carrying an electric current, and (2) these forces are always directed in accordance with a simple rule whatever the actual positions of the magnet and the wire. In fact, several rules have been devised for remembering the direction the force will try to move the two relative to one another. One such rule is: Imagine you are swimming along the wire in the direction of the electric current and looking at the N. pole of the magnet, then that N. pole will tend to move in the direction of your outstretched left hand.

The forces we are considering are present because a magnetic field is created around a wire carrying an electric current. We do not know why this magnetic field exists but we know that it does and we make use of it in many ways, as we shall see later in this chapter.

Look at the directions of the magnetic lines of force as traced on the cards shown in Figure 28.2 (c) and (d). The magnetic lines of force of the coil or solenoid enter at one end and leave at the other in exactly the same way that they would in the case of a magnet and thus they give to the coil or solenoid the same magnetic properties as a real magnet. For example, look at the near face of the coil drawn in Figure 28.2 (c). The current there is seen to flow in a clockwise direction and there is a 'S. pole' at this face because the magnetic lines of force enter through this end of the coil. There is a 'N. pole' at the face on the other side of the coil.

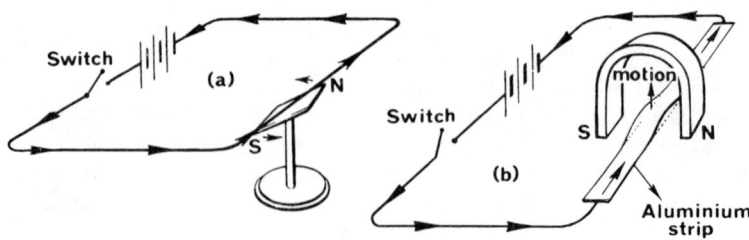

Figure 28.1. The relative movements of a magnet and a wire carrying an electric current when (a) the magnet is free to move but not the wire, and (b) the wire (aluminium strip) is free to move but not the heavy horseshoe magnet.

The magnetic field produced in a solenoid is very much stronger when an iron core is placed inside it along its axis. This combination of solenoid and core is then known as an *electromagnet*. There are many different ways in which electromagnets are designed—their shapes and strengths vary according to the work they have to perform. Most powerful electromagnets have very many turns of wire wound closely together to form a coil around the iron core.

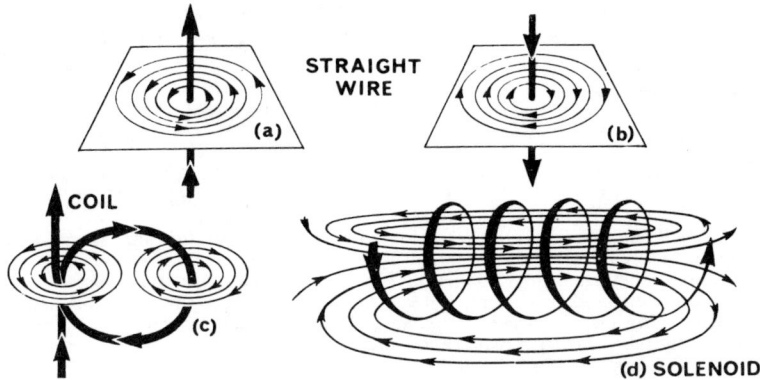

Figure 28.2. The magnetic fields produced by an electric current flowing along a wire. The thick arrows show the directions of the electric current in the wires, and the thin arrows the direction of the magnetic lines of force surrounding the wires.

Figure 28.3. The end faces of a coil or solenoid looked at from a position along its axis and away from it. The arrows help us to remember the polarity of the magnet formed when the current flows in the directions indicated.

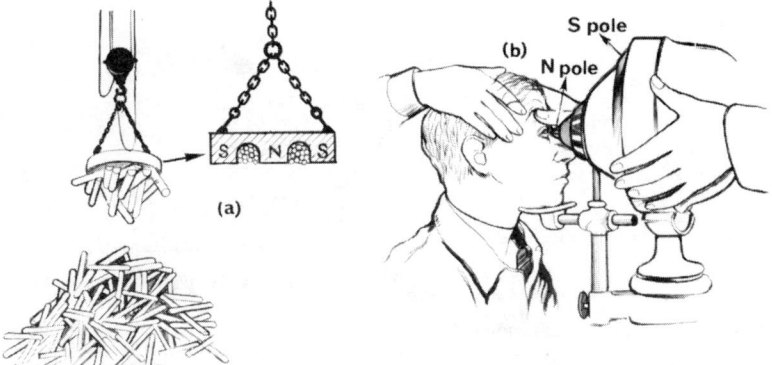

Figure 28.4. Powerful electromagnets used (*a*) to lift scrap iron, and (*b*) to extract an iron filing from an eye.

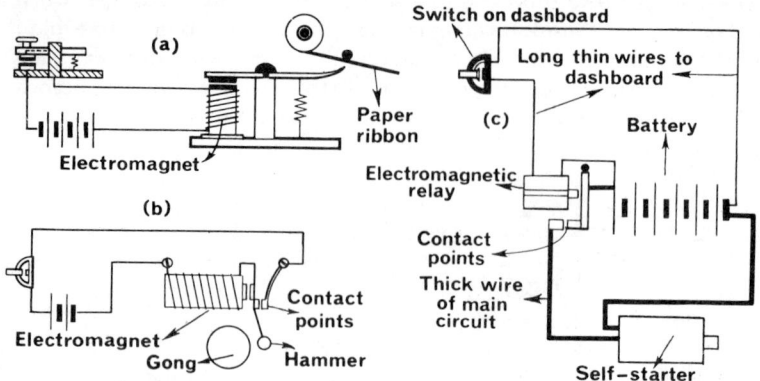

Figure 28.5. Some uses of an electromagnet. (*a*) A key starts and stops the electric current flowing in the circuit of the telegraph and thus enables a message to be transmitted and tapped out in code on a paper ribbon. (*b*) A vibrating spring and armature causes a regular interruption of the electric current through the electric bell and thus produces a steady tapping of the hammer on the gong. (*c*) A small electric current passing through the electromagnet of this relay closes the main circuit and causes a large electric current to flow from the battery through the self-starter of the motor-car engine.

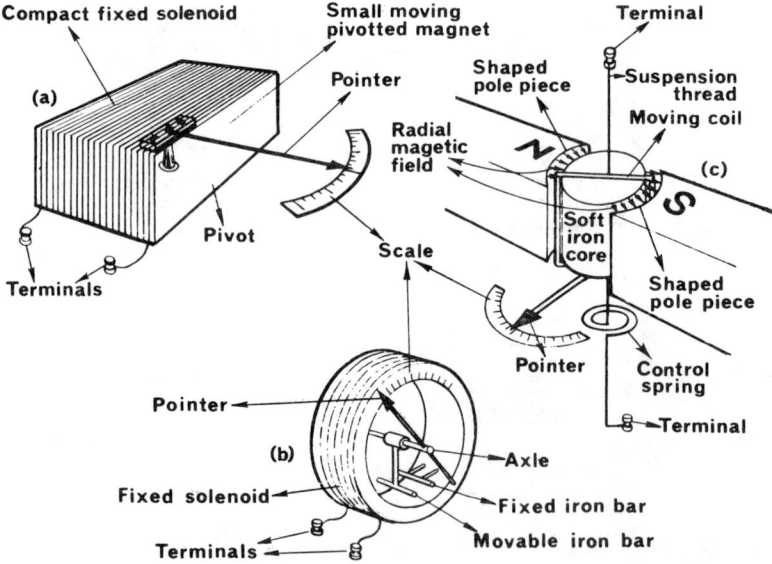

Figure 28.6. Types of galvanometers. (*a*) A moving magnet. (*b*) A moving iron. (*c*) A moving coil in which shaped pole pieces and a central soft iron core ensure that the moving coil turns in a strong uniform radial magnetic field. What control is applied in each of the types illustrated to prevent the smallest current from swinging the pointer the full distance across the scale?

If the iron core of an electromagnet is made of very soft iron then as soon as the electric current ceases to flow through the coil the magnetism disappears. Because the force of attraction between the electromagnet and an iron armature or disc is created and destroyed easily and quickly, we are able to design many useful circuits involving switching mechanisms.

Galvanometers are instruments used to detect and sometimes to measure very minute electric currents. There are three main types of these instruments and all of them depend for their action on the magnetic effects of these small electric currents. One is the moving magnet type in which a small magnet is made to twist in the magnetic field produced in a coil of wire. Another, the moving iron type, is dependent upon the repulsion between two iron bars lying side by side in a solenoid both of which become magnetized. The third, known as the moving coil type, has a fixed magnet inside which moves a small coil. The magnetic field created by the passing of the electric current through the coil reacts with the magnetic field of the fixed magnet and provides the couple that twists the coil against its control spring and suspension thread. The most sensitive of the three instruments is the moving coil type; most galvanometers are constructed with a moving coil movement. The moving iron type is the only one of the three that can detect both direct and alternating current.

These galvanometers are the basic meter movements used in the construction of current measuring instruments called *ammeters*, and potential measuring instruments called *voltmeters*.

Ammeters are designed to measure larger currents than galvanometers measure. They are made to do this by arranging for some of the current to travel along a parallel resistor, called a *shunt*, leaving only a small fraction to pass through the basic current-measuring galvanometer. See Figure 28.7(a). This small fraction is indicated by the pointer on a scale that is calibrated to show the value of the whole current through the galvanometer and the shunt combined. The only difference between ammeters measuring maximum

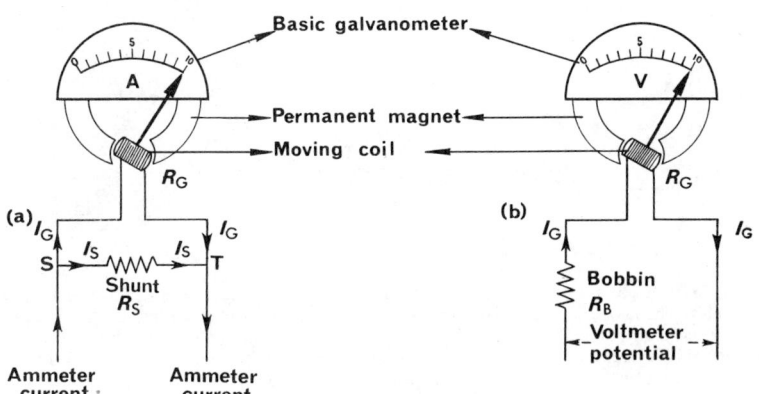

Figure 28.7. The same basic current-measuring galvanometer connected (*a*) to a shunt to convert it into an ammeter, and (*b*) to a bobbin to convert it into a voltmeter.

currents of 1 ampere, 5 amperes, or 10 amperes is in the value of the shunts attached to the same basic galvanometer. Hence some ammeters combine a shunt and its appropriate scale together so that they can be 'plugged in' at the same time. The shunt is thus connected across the terminals of the galvanometer at the same time as the correct scale is placed in position. See Figure 28.8

How is the resistance of the *shunt* needed for a particular ammeter calculated when the details of the galvanometer are known? It is clear from Figure 28.7 that the potential between the points S and T is the same whichever way the current travels—either through the galvanometer or through the shunt. Suppose the galvanometer has a resistance of 5 ohms and the current needed for a full-scale deflection is 0.015 amperes. What is the value of the resistance of the shunt required to make the ammeter measure up to 10 amperes?

Applying the Ohm's law equation to the galvanometer branch of the circuit:

Potential between S and T, $\qquad V_G = I_G \times R_G$

$$\text{or} \quad V_G = 0.015 \times 5$$

$$\therefore \ \underline{V_G = 0.075.}$$

Now apply it to the shunt branch of the circuit and remembering that $V_G = V_S$, and that I_S must be $(10 \text{ amp} - I_G)$ i.e. 9.975 amp:

Potential between S and T, $\qquad V_S = I_S \times R_S,$

$$\text{or } 0.075 = 9.975 \times R_S$$

$$\therefore \ \underline{R_S = 0.007\ 55}$$

Therefore to convert the galvanometer to an ammeter with a full-scale reading of 10 amperes a shunt of 0.007 55 ohms must be placed in parallel with it. By making similar calculations the value of the shunt needed to convert any galvanometer into an ammeter to read any chosen maximum current can be determined.

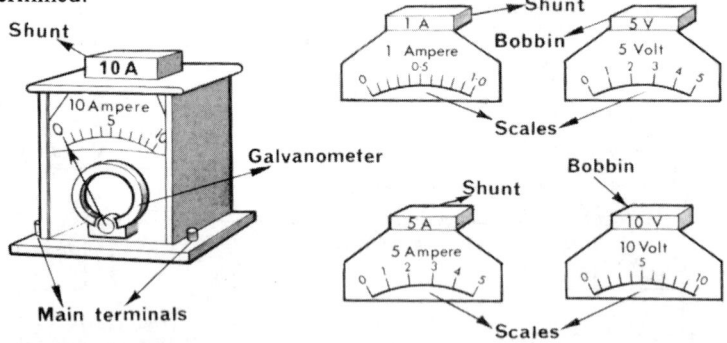

Figure 28.8. A galvanometer that can be adapted easily to function as an ammeter or as a voltmeter. The instrument as shown in the diagram is an ammeter reading up to a maximum current of 10 amperes. Other combined shunts and scales shown convert the galvanometer to give different current readings when inserted in place of the 10 ampere combination. Combined bobbins and scales are also shown. When one of them is plugged into the galvanometer (using different plugs and sockets) the instrument is converted into a voltmeter with the selected maximum scale reading of potential difference. By interchanging resistors and scales in this convenient way you can use one expensive galvanometer movement for many purposes.

Voltmeters measure the e.m.f. or potential difference (i.e. voltage) between two points. The basic galvanometer, which was described in the calculations above, could be used without modification as a voltmeter giving a full-scale deflection when the voltage is 0.075 volts. This same galvanometer can be used as the basic measuring device in a voltmeter to read any chosen voltage, provided that a resistor is placed in series with it to keep the maximum potential difference across the galvanometer at this value of 0.075 volts whatever the potential difference is across the whole instrument. The series resistor is often called a *bobbin* because the long length of wire forming the resistor is wound on a bobbin and simply attached to one terminal of the galvanometer to convert it into a voltmeter. See Figure 28.7.

How is the resistance of the bobbin needed for a particular voltmeter calculated? Let us use, as an example, the same basic galvanometer that we described in the calculation above, and consider how to convert it into a voltmeter which will have for its maximum reading 10 volts.

Applying the Ohm's law equation to the bobbin and to the galvanometer separately:

The current passing through the bobbin and the galvanometer is the same,

$$\text{or } I_\text{B} = I_\text{G} = 0.015 \text{ A}$$

The potential difference across the bobbin must be made 10 volts less the potential difference across the galvanometer, or

$$V_\text{B} = (10 - 0.075)$$

$$\therefore \ \underline{V_\text{B} = 9.925.}$$

Applying the Ohm's law equation to the bobbin:

$$R_\text{B} = \frac{V_\text{B}}{I_\text{B}}$$

$$\therefore \ R_\text{B} = \frac{9.925}{0.015}$$

$$\therefore \ \underline{R_\text{B} = 661.6}$$

Therefore to convert the galvanometer to a voltmeter with a maximum reading of 10 volts a bobbin of 661.6 ohms must be placed in series with it. By making similar calculations the value of the bobbin needed to convert any galvanometer into a voltmeter reading to any maximum value can be determined.

There is one thing the addition of a shunt or a bobbin to a galvanometer cannot do; that is to convert it to a more sensitive instrument reading smaller values of current or potential difference.

The loudspeaker is another practical application of the magnetic effects of an electric current. In many ways it is similar in construction to the galvanometer. The speech current passes through a light movable coil made of a few turns of wire or metal tape suspended inside a strong fixed magnetic field. The coil has attached to it a light fabric cone that is usually skilfully fixed at its wider end to a rigid framework in such a way that it can vibrate to reproduce the required sounds. The variable magnetic field produced by the varying electric current in the coil reacts with the fixed magnetic field thus causing the coil to vibrate in sympathy with the varying currents passed through it. The fabric cone firmly attached to the coil also vibrates and emits sound waves in the air. In this way a loudspeaker converts the energy supplied by a pulsating or varying electric current into sound waves. There are two ways of producing the fixed magnetic field—one is by using a permanent magnet as in the galvanometer, and the other is by using a powerful electromagnet energized by some external supply of direct current electricity.

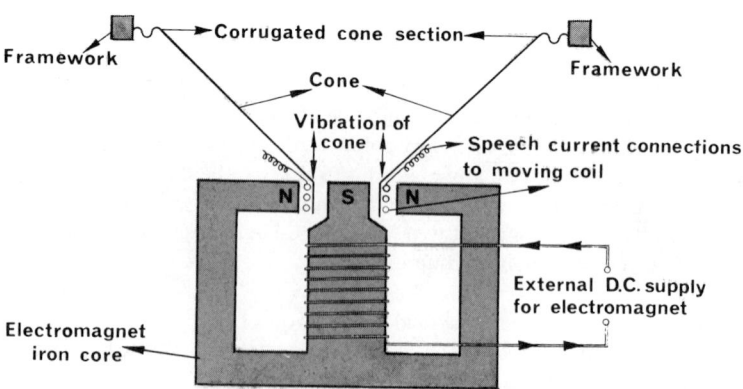

Figure 28.9. A loudspeaker. The moving coil attached to a light cone vibrates between the powerful magnetic poles of an electromagnet.

Questions on chapter 28

1. How can you show that a magnetic field exists around a conductor carrying a direct current of electricity?

2. Draw a diagram of an electromagnet and state two ways of increasing its strength.

3. If you were making an electromagnet what three things would determine its strength when you used it? Why is soft iron used rather than steel in the electromagnet of an electric bell?

4. Draw a diagram to show how you would use an electromagnet to magnetize a steel bar. Mark in all the poles.

5. Why does the hammer of an electric bell return after striking the gong? Draw a diagram of an electric bell and show clearly the electrical circuit.

6. Explain the action of an electromagnetic relay. Illustrate your answer with a clear diagram. What are the uses of a relay?

7. What advantages have electromagnets over permanent magnets? In what practical devices are electromagnets used?

8. When is it desirable to use a horseshoe electromagnet instead of a straight electromagnet? In what practical devices are horseshoe magnets used?

9. Describe by means of diagrams the construction of a moving coil galvanometer. How could you convert this galvanometer into (a) an ammeter, and (b) a voltmeter?

10. A galvanometer coil has a resistance of 20 ohms. The pointer moves over an even scale and registers a maximum reading when 0.01 ampere is passing through it. Calculate (a) the resistance of the shunt that must be attached to it to convert the galvanometer into an ammeter whose maximum scale reading is 1 ampere, and (b) the resistance of the bobbin that must be connected in series with it to convert the galvanometer into a voltmeter whose maximum scale reading is 15 volts.

29. The heating and lighting effects of an electrical current

When electricity is passed through wires in a circuit much of the energy of the moving electrons is turned into heat as the electrons are forced to flow against the resistance offered by the wires. Sometimes a device using this principle, if properly controlled and designed, is a convenient method of obtaining heat, but sometimes it can be dangerous and create fires, as in badly and inadequately wired buildings. It is the most commonly used process in the conversion of electrical energy into heat energy.

Many electrical devices in the home can be made to provide heat in just the correct position and quantity by choosing suitable coils of wire. Most wires are made of alloys because they have the properties most desired for this purpose. The alloy of nickel and chromium, called *nichrome*, has a high resistance, a high melting point, and it is not easily oxidized by air, so that it can become red hot and remain so without burning away. Some wires are wound in fine spirals and others are flattened into ribbons, according to where they are to be used. Some of the devices incorporate in their circuits, in series with the heating elements, a bimetallic strip *thermostat* which automatically switches off the electric current when a predetermined temperature is reached and switches it on again when the temperature falls.

It is essential in any wiring system to keep the heating elements where they are wanted and not to introduce accidentally in the connecting wires conditions that could produce excessive heat which might set fire to inflammable material near by. One must not overload the house circuits by connecting appliances that draw more current than the connecting wires are designed

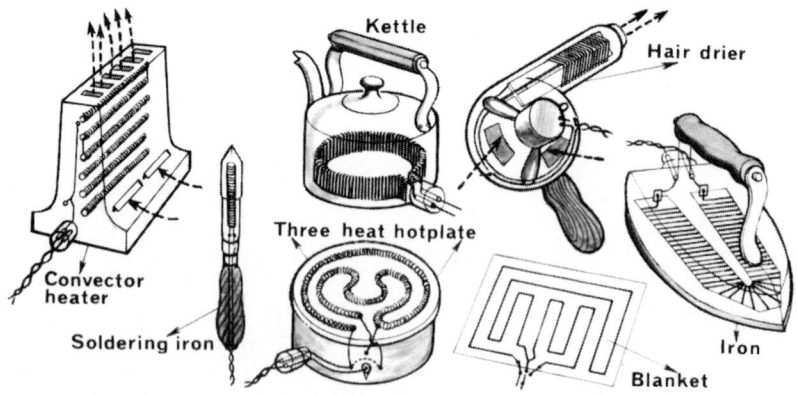

Figure 29.1. Various types of heating elements are shown in these electrical appliances. Why are the heating elements placed in the positions shown? What electrical power do you think each one consumes?

to carry. For example, an electric fire must not be connected to an electric lamp socket. The connecting wires joining an electric appliance to its plug must not be allowed to be strained or worn so that the current can flow directly across them through broken insulation, forming a *short circuit* and producing an excessive current flow.

All electrical appliances should be 'earthed' so that if the metal connecting wires do touch the body of the appliance, either through frayed ends or cracked insulation, a short circuit would occur direct to 'earth' and not through the person who is touching the appliance. The earth connection to the appliance is made through a third wire joined to a third pin in the plug, which connects it through the socket to a metal rod driven into the real earth outside or to an earthed metal water pipe.

Fuses protect buildings from the hazards of the overheating of the conducting wires, short circuits, and the overloading of electrical appliances. A fuse is a short length of wire, usually an alloy, that easily melts at a low temperature. It is placed in the circuit so that the current flowing must pass through it. The thickness of the fuse wire is so chosen that if the current exceeds the maximum for which the wiring system is designed the heat developed in the fuse melts it and thus breaks or opens the circuit. When a fuse 'blows' a small electric arc is created at the gap so formed and the hot molten metal is scattered. Thus it is usual to encase the fuse wire in a fireproof container to prevent it setting fire to any inflammable material. A blown fuse should always be replaced by another of the same maximum current-carrying capacity—*never by a bigger one*. If the replacement fuse 'blows' it is a certain sign that the circuit is overloaded somewhere and the cause of the overloading should be found and corrected before the fuse is replaced again. In power stations where very large

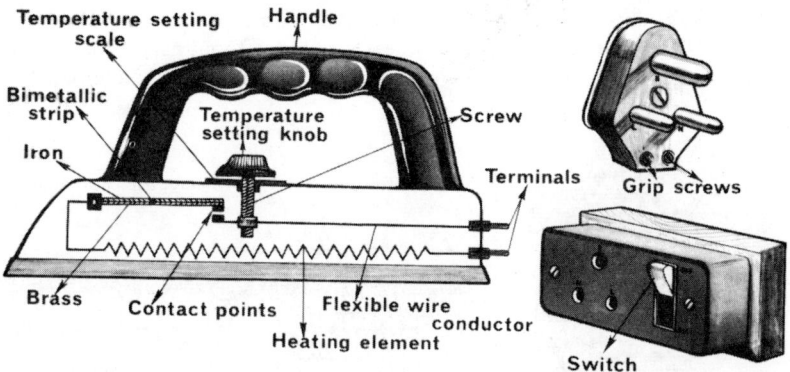

Figure 29.2. A domestic electric iron. Its temperature is determined by the position of the knob and screw. When that temperature is reached the bimetallic strip bends upwards and opens the contact points. When the iron cools the bimetallic strip closes the contact points again and the current passes once more through the heating element. In this way the thermostat maintains a selected temperature with only slight variations.

Figure 29.3. A standard three-pin plug and switch socket. E is connected to the green and yellow earth wire, L to the brown live wire, and N to the blue neutral wire. The cable is secured by tightening the grip screws. A variety of shapes of pins exist but the general arrangement is the same.

currents are controlled fuses are too dangerous to use because of the large mass of the hot molten metal formed when the fuse blows and the electric arcs that may be formed. Large electromagnetic circuit breakers are therefore employed.

Electric heating is used in many industrial processes because it can be so easily controlled.

One type of electric *furnace* gets hot because of the *resistance* offered to the electric current either by the material itself or by carbon rods around which the material to be heated is packed. Another type is the electric *arc* furnace in which the material, placed in a large vessel, is heated by an arc formed by carbon rods. Such furnaces develop temperatures up to 3 500° C.

Electric welding is a method of joining two pieces of metal together. '*Resistance*' welding employs a large electric current and a great pressure at the contact points to be welded. This is sometimes called 'spot' welding because two sheets of metal can be fastened together at a number of selected spots instead of all over an area. Another method is *arc* welding by which the heat for welding is derived from an arc formed between the base metal and an electrode of filler metal.

Electric lamps contain heating elements (filaments) that become so hot when an electric current flows through them that they give off light. The present electric lamp has a high-resistance tungsten filament enclosed in a glass bulb filled with a mixture of two inert gases, argon and nitrogen. The filament is a coiled coil of fine tungsten wire that becomes brilliantly incandescent without burning or melting. The story of the development of the electric lamp tells how scientists by their patient labours have overcome many difficulties to produce the robust lamp of today. They have still one more problem to solve and that is to produce a much more efficient lamp—one which gives light without wasting as heat energy over 90% of the electrical energy supplied to it.

The long bright tubular *fluorescent lamp* that works on an entirely different principle is almost the answer to the quest for light without heat. This lamp is filled with mercury vapour and is coated on the inner surface with a fluorescent

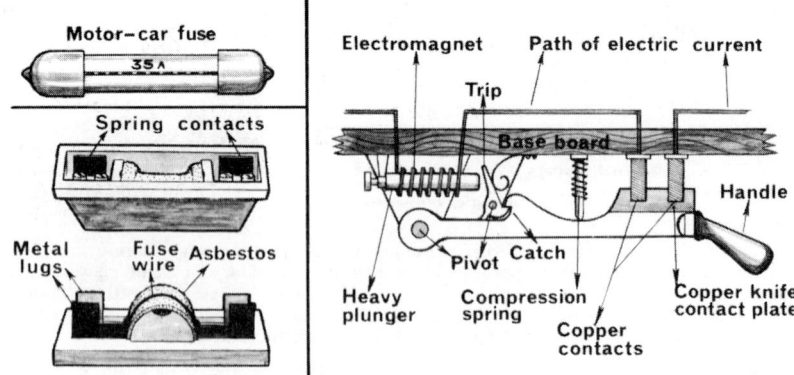

Figure 29.4. Two fuses suitable for use with small electric currents, and an electromagnetic circuit breaker for large currents.

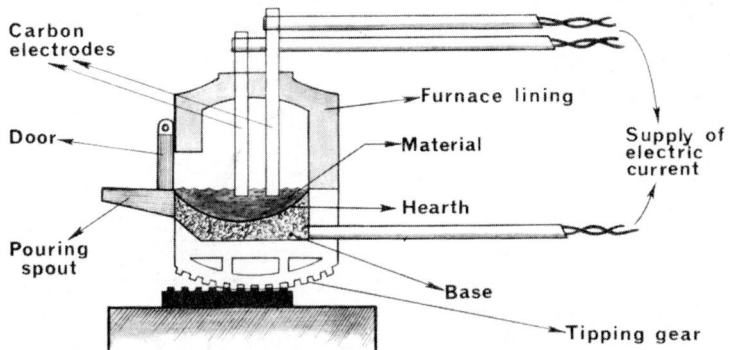

Figure 29.5. An electric arc furnace. The arc is formed between the carbon rods and the base of the furnace so that the current passes through the material in the hearth.

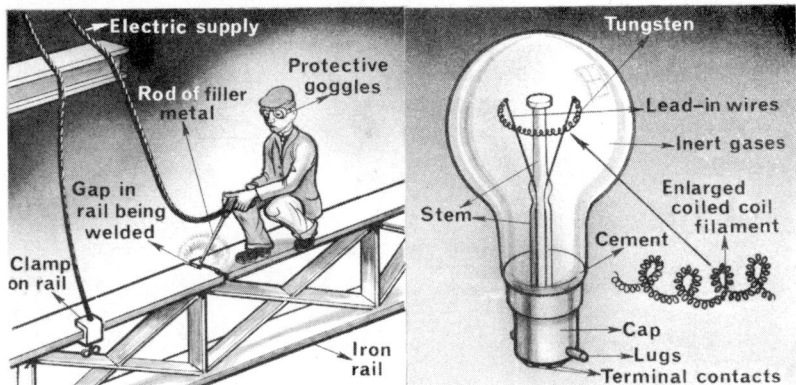

Figure 29.6. Electric arc welding. A rod of 'filler' metal forms an arc with the base metal plate. The operator protects his eyes from the intense glare of the arc and an excess of ultra-violet light by wearing dense purple glasses through which he can see the hot metal and the welding process.

Figure 29.7. A coiled coil filament electric lamp. The wire is made into a tight coil and then this is wound into a bigger coil.

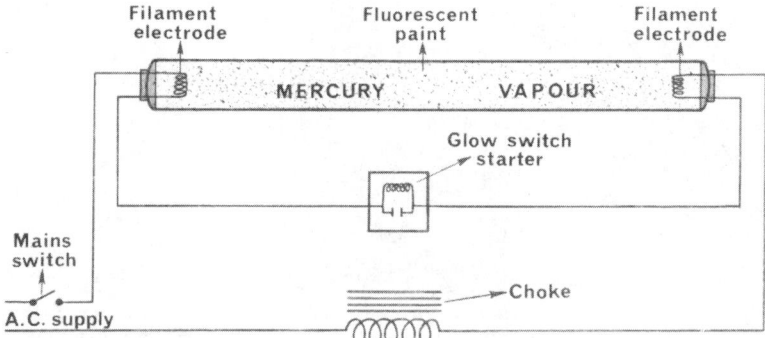

Figure 29.8. A fluorescent lamp. At first the circuit is completed through the two filaments and the starter. After a short interval the starter opens. Then a sudden surge of voltage occurs between the two filament electrodes caused by the interruption of the current through the choke. This voltage surge creates a discharge in the mercury vapour in the tube. The mercury vapour emits ultra-violet rays that cause the paint to fluoresce.

P

paint that glows when the ultra-violet light emitted by the mercury vapour strikes it. The colour of the light given off outside the tube depends on the composition of the fluorescent paint.

Electricity and heat

We have observed that heat energy is produced by an electric current only when electrical energy is expended. We have learned also that energy cannot be created or destroyed. We are therefore not surprised to learn that a definite relationship exists between heat and electrical energy. It was James Prescott Joule who first established this 'electrical equivalent of heat'. He passed an electric current through a coil of wire immersed in a calorimeter and observed the rise in temperature of a certain mass of water in a certain time. He knew also what current was flowing through the wire and the potential difference across its ends. From his observations Joule stated that '*The amount of heat produced in any conductor is directly proportional to the square of the current, the resistance of the conductor, and the time during which the current flows* '

The laboratory experiment shown in Figure 29.9 is a simplified version of the experiment performed by Joule. This experiment gave an average temperature rise of 16.5° C. Thus the heat gained by the water can be calculated by the equation first stated on page 139 namely

$$Q = m \times (T_1 - T_2) \times c$$

where m is the mass of the water (in kg), $(T_1 - T_2)$ is the change in temperature (in °C), and c is the specific heat capacity (in J kg^{-1} °C^{-1}).

Hence the heat gained = 0.4 × 16.5 × 4 200 J,
= 27 720 J.

The electrical energy expended is the product of the power (in W) and the time (in s).

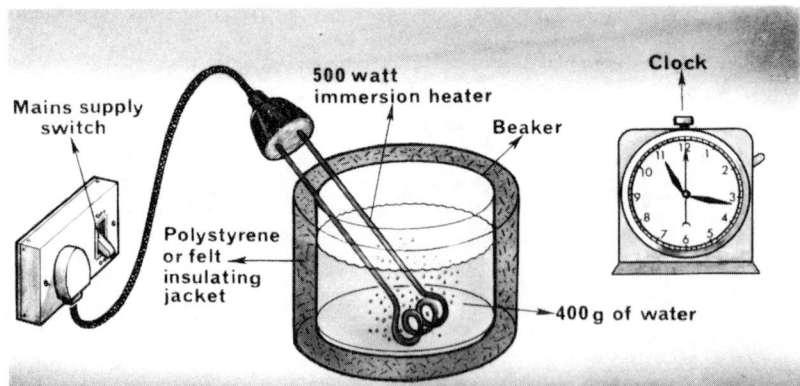

Figure 29.9. How to set up a laboratory experiment to determine the electrical equivalent of heat. The 500-watt immersion heater is switched on for 60 seconds and the rise in temperature of the 400 grams of water is noted. The average reading of several experiments is observed.

Hence the electrical energy expended $= 500 \times 60$ watt-seconds
$$= 30\,000 \text{ watt-seconds.}$$

If we examine these two values they do not equate exactly as we should expect from the statements on page 194 where 1 joule of energy is provided by 1 watt for 1 second. Why do you think that the heat measured in the beaker is less than that calculated electrically? Some of the 30 000 watt-seconds or joules have been lost because only 27 720 joules were registered in the beaker.

Joule's laws of electric heating can be expressed in the form of equations:

$$Q = I^2 \times R \times t$$

or
$$Q = I \times V \times t$$

or
$$Q = \frac{V^2}{R} \times t$$

These equations are all the same as you can verify by applying the fundamental relationship expressed by Ohm's Law as on page 196,

$Q =$ the quantity of heat in joules,
$I =$ the current in amperes,
$V =$ the voltage in volts,
$R =$ the resistance in ohms
$t =$ the time in seconds.

A joule or watt-second is a small unit of energy and for practical purposes we buy electrical energy in kilowatt-hours (kWh) where
$$1 \text{ kWh} = 1\,000 \times W \times 60 \times 60\,\text{s} = 3.6 \times 10^6 \text{ joules}$$

Example 1. How long will it take a kilowatt electric kettle to bring 3 litres of water at 28° C to the boiling point of water?

Heat required to boil the water

$\quad\quad = $ mass of water $\times$ specific heat capacity of water
$\quad\quad\quad\quad\quad\quad \times$ rise in temperature of water,
$\quad\quad = m \times (T_1 - T_2) \times c$
$\quad\quad = 3 \times (100 - 28) \times 4\,200 \text{ J}$

$\therefore$ Time needed $= \dfrac{3 \times 72 \times 4\,200}{1000}\,\text{s}$

$\quad\quad = \underline{15.1 \text{ min}}$

$\therefore$ it will take 15.1 minutes to boil the 3 litres of water

Example 2. What will it cost to use the following electrical appliances for 2 hours assuming that a unit of electricity (1 kWh) costs $1\frac{1}{2}$ new pence:

(a) 100 watt lamp,
(b) 2 kilowatt fire,
(c) 15 watt glow lamp,
(d) an electric fan drawing a current of 1 ampere from a 240 volt supply?

To solve those questions we shall calculate (1) the electrical power taken in watts, (2) the electrical energy supplied in kilowatt-hours, and (3) the cost at $1\frac{1}{2}$ new pence per kilowatt-hour. The answers are tabulated thus:

Appliance	Power in watts	Energy in kWh	Cost in new pence
a	100	0.2	0.3
b	2 000	4.0	6.0
c	15	0.03	0.045
d	240	0.48	0.72

Heat control. Many electrical hot-plates, blankets, and other heaters are designed to supply three different heats usually marked on the switch as LOW, MEDIUM, and HIGH. There is also an OFF position. These heaters are constructed with two equal coils of high resistance wire in series which form the heater elements. The switch is complicated but you can understand how it operates from the four wiring diagrams illustrated in Figure 29.10.

In the OFF position one terminal, connected to the mains supply, is disconnected entirely from both heater elements and thus no current flows at all. The hot-plate therefore remains cold.

In the LOW position the current flows through both the elements in series. Using the values shown in the wiring diagrams the resistance of the elements in series is 60 + 60 = 120 ohms. Therefore the power consumed by the hot-plate in the LOW position can be calculated as follows:

$$\text{Power} = \frac{V^2}{R} \text{ watts,}$$

$$\therefore \text{Power} = \frac{240 \times 240}{120} \text{ watts,}$$

$\therefore$ Power in LOW position = 480 watts.

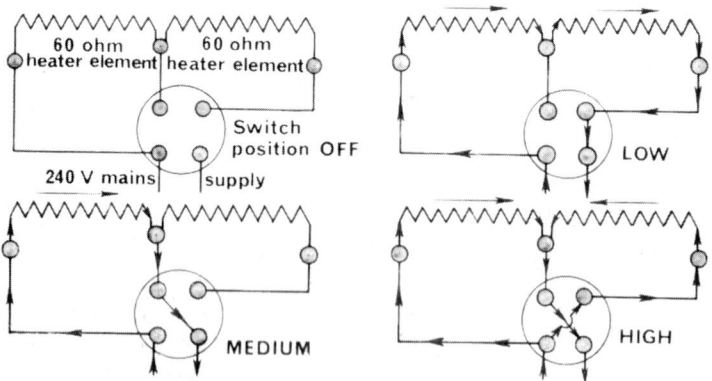

Figure 29.10. The wiring diagrams of an electrical hot-plate and a three-heat control switch shown in the four positions of OFF, LOW, MEDIUM, and HIGH. This particular hot-plate operates from a 240 volt supply and has two heater elements of 60 ohms each in series. In order to make the circuits in use clear they have been indicated by arrows.

In the MEDIUM position there is one element only in the circuit. This has a resistance of 60 ohms. The power consumed is calculated in the same way as before:

$$\text{Power} = \frac{240 \times 240}{60} \text{ watts,}$$

∴ Power in MEDIUM position = 960 watts.

In the HIGH position the current passes along two circuits and each has a resistance of 60 ohms. The power consumed is the sum of the powers consumed in each circuit and is calculated as follows:

$$\text{Power} = \frac{240 \times 240}{60} + \frac{240 \times 240}{60} \text{ watts,}$$

∴ Power in HIGH position = 1920 watts.

The heat output of the hot-plate is, in each case, supplied by the power consumed by the heater elements. In other words the heat outputs are in the ratio of 1, 2, and 4 for the LOW, MEDIUM, and HIGH positions of the control switch.

Questions on chapter 29

1. A house has an a.c. supply of 220 volts. You are presented with an electric fire containing two elements each of 750 watts. The fire contains a switch to use two elements or one. Give a diagram to show how the two elements and the switch are connected.

2. Draw diagrams to show how two heating elements may be connected to a given supply to give HIGH and LOW powers.

3. Explain with the aid of diagrams the action of some form of thermostat for controlling the heat supplied by some form of electric heater.

4. A 2kW electric fire is operated on a 200 V supply.

What is (a) the current taken, (b) the resistance of the fire, and (c) the cost in pence of running it for 10 hours if the cost of electricity is $1\frac{1}{2}$ new pence a unit?

5. Draw a diagram of the electrical installation in a house to supply two power points and a lighting circuit.

Show clearly the earths provided and the positions of all fuses.

What is the cost in new pence of using a 3 kW heater five hours a day for a week if the cost of the electricity is $1\frac{1}{2}$ new pence a unit?

6. Draw a diagram of the wiring necessary to enable a person to switch a light on or off from either of two places. What type of switches are used in this circuit?

An electric fire rated at 200 V 1 kW is connected to a 200 V supply. What current will flow? What is the resistance of the heating element of the fire when it is hot?

7. Describe and explain, with the aid of a diagram, the construction and action of an electrical fuse.

A house with a 240 volt mains supply on one 5 A circuit contains a 500 W electric fire, two 100 W lamps, and five 60 W lamps. What would be the cost in new pence per hour of running the full electrical load of this circuit in the house, if electricity costs 5 new pence per unit? How many more 60 W lamps could be installed in the same circuit without burning out the 5 A fuse?

8. What is measured by (a) a watt, and (b) a kilowatt-hour?

An electric fire is marked '250 volts, 500 watts'.

Calculate (c) the current it takes when connected to the 250 volt mains supply, and (d) its resistance. What would be the weekly cost in new pence of using this fire for two hours a day, if electricity costs 4 new pence per unit?

9. An electric fire rated at 500 W is connected to a 250 V supply. Calculate (a) the current flowing through the element, (b) the number of units of electricity used in 5 hours, and (c) the cost in new pence of using the fire for a week if it is switched on for 5 hours each day, the electricity costing 6 new pence per unit.

10. Two electric fires, one of 750 W and the other of 1 000 W, are connected in parallel to a 250 V supply. Calculate (a) the total current taken when both are switched on, (b) the number of units of electricity used in 6 hours when both fires are switched on, (c) the daily cost in new pence of using them if the first fire is used for 3 hours a day and the second fire for 8 hours a day, electricity costing 6 new pence per unit, and (d) the maximum number of electric lamps each taking a power of 200 W that could be used on this 250 V supply if nothing else is switched on and the circuit has a 13 A fuse in it.

30. The chemical effects of an electric current

We have already studied how chemical energy as it exists in a primary cell (Leclanché) is transformed into electrical energy when the cell is in use. We have also learned that chemical energy and electrical energy are interchangeable in the case of the processes of charging and discharging a secondary cell (accumulator).

Electrolysis. If a supply of d.c. electricity is allowed to pass through a solution of certain chemical compounds it produces changes in the solution, or in some cases it breaks up the chemical compound and separates its different parts. This process is known as electrolysis.

All soluble acids, salts, and bases form *electrolytes* or conducting solutions when dissolved in water. Typical examples of acids are hydrochloric and sulphuric acids, typical salts are copper sulphate and sodium chloride (common salt), and two common bases are sodium hydroxide (caustic soda) and ammonium hydroxide (household ammonia). When these compounds are dissolved in water their separate atoms or groups of atoms (e.g. SO_4) gain or lose electrons and form *ions*. For example, in a water solution sodium chloride forms a positively charged sodium ion (written Na^+) and also a negatively charged chlorine ion (written Cl^-). This means that the sodium atom gives up one electron and the chlorine atom receives one electron when the salt splits up in water. Obviously one electron is transferred from the sodium atom to the chlorine atom.

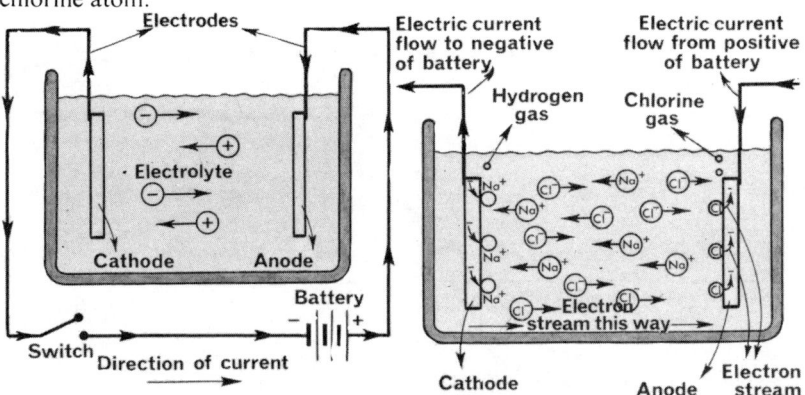

Figure 30.1. An electrolytic cell in operation. The electric current passes from the anode to the cathode through the electrolyte. The direction of the charged ions is shown by the arrows. The electrons travel from the cathode and are released at the anode.

Figure 30.2. The electrolysis of a solution of sodium chloride using carbon electrodes. Sodium hydroxide is formed in the solution with the liberation of chlorine bubbles at the anode and hydrogen bubbles at the cathode.

221

In the process of the electrolysis of sodium chloride the chlorine ion Cl^-, because it is negatively charged, moves to the anode, gives up its electron there and combines with other uncharged chlorine atoms to form bubbles of chlorine gas that escape to the surface. This is the direction that electrons travel in the whole circuit through the battery and the electrolytic cell. The Na^+ ion moves in the opposite direction to the cathode. Water also ionises to a very limited extent to form hydrogen ions H^+ and hydroxide ions OH^-. The Na^+ ions and the OH^- ions form sodium hydroxide (caustic soda). The H^+ ions attract electrons⁻ from the cathode and form H atoms which then combine in pairs to rise as bubbles of H_2 molecules.

From the above description of how electrolysis takes place it is clear that a certain fixed quantity of an element (one atom in the case of chlorine) is liberated for every electron that flows through the circuit. Hence the quantity of an element liberated by electrolysis is proportional to the quantity of electricity passed through the electrolyte. This is an important relationship because as we shall learn later in this chapter it will enable us to calculate how much of a pure substance can be extracted from an impure one, and how thick a layer of one element can be deposited on another when a certain quantity of electricity is used.

Electroplating is one special application of electrolysis. It is the process of coating with a layer of a particular element an object suspended as the cathode in an electrolytic cell. For example, the iron bumpers and radiators of motor-cars are coated in this way first with nickel and then with chromium to prevent rust forming on them. Baser metals are often coated with other metals to improve their appearance. Thus we electroplate brass spoons with nickel or silver, and articles made of some gold alloys with a very thin coating of pure gold. To electroplate an article it is only necessary to make the article the cathode, choose a suitable electrolyte that will furnish the ions of the plating metal, and make the anode of a bar of the plating metal.

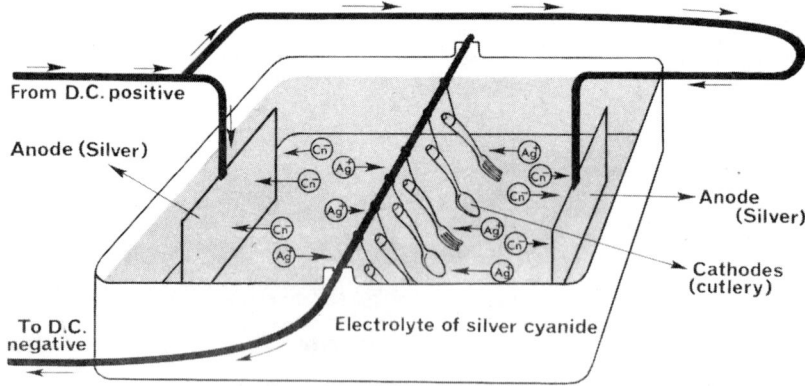

From D.C. positive

Anode (Silver)

Anode (Silver)

Cathodes (cutlery)

To D.C. negative

Electrolyte of silver cyanide

Figure 30.3. The electroplating of brass tableware with a coating of silver can be considered as an electrolysis of a solution of silver cyanide in water. The commercial silver plating process uses more complicated electrolytes but the same method.

Refining of metals. The process of electroplating is also used to *refine* certain metals by coating layer upon layer of the pure metal on a cathode made of the pure metal. For example, copper, gold, silver, zinc, and tin are all separated from their impurities by this means.

Some metals such as aluminium, magnesium, calcium, sodium, potassium, and rhodium are extracted directly from their ores by electrolysis. North of the Great Lakes in Canada impure boulder (or native) copper is found in the earth. An ingenious electrolytic method is used for its extraction and purification. An anode is drilled into the impure copper rock, several pure copper cathodes are inserted into an artificial lake or sometimes into a natural lake if one exists, and some copper sulphate is thrown into the lake as an electrolyte. A strong electric current is passed between the electrodes so that electrolysis proceeds with the lake acting as a huge electrolytic cell. The enlarged copper cathodes are withdrawn and are replaced by thin ones. The boulder copper slowly disappears leaving behind only earth and rocks as sludge.

Electrotyping is one of the methods used by printers for making a permanent record of a book from which further printings may be made. Ordinary type is set up and a wax impression made of it. The wax is then coated with fine graphite to make a conducting layer on to which copper is electroplated in the normal way. The wax is removed, the copper impression is strengthened with a backing of a strong metal alloy, and then these 'electroplates' are ready for use on printing machines.

Gramophone records are prepared, by a similar process, from wax discs on which wavy grooves have been drawn by a stylus tracing the original sound waves of the recorder. The wax disc is covered with fine graphite, the copper impression is made by electroplating, the copper shell is strengthened, and finally warm plastic material is pressed against the copper disc by a hydraulic press. When the plastic cools it becomes hard and firm enough to enable sound to be reproduced from the movements of a fine needle following along the wavy grooves.

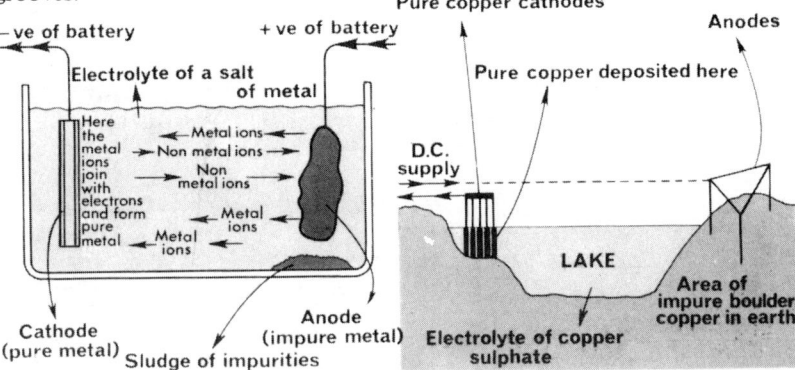

Figure 30.4. The refining of metals by electrolysis. The cathode is a thin sheet of the pure metal, and the anode is a large plate or mass of the impure metal.

Figure 30.5. The extraction and purification of copper directly from large-scale deposits found in the earth.

Questions on chapter 30

1. Draw a diagram of the circuit used in electroplating including a switch, something to control, and something to measure the current used. Mark clearly the positive and negative poles of the electrical supply point and the position of the object to be plated. Name all the essential parts.

2. Draw a labelled sketch of apparatus suitable for carrying out the electrolysis of water. State the products formed and their proportions.

3. A metal spoon is to be coated by electrolysis with a layer of copper. Draw and label a circuit diagram of the apparatus you would use. Also mark on the diagram the anode and the name of the electrolyte. Indicate the value of the current required.

31. Electromagnetic induction

We have already learned of the discovery, made in 1819 by the Danish professor Hans Christian Oersted, that a magnetic needle is influenced by the presence of an electric current flowing in a wire. After this discovery physicists in many countries argued that if an electric current could produce a magnetic field somehow or other a magnetic field ought to be able to produce an electric current.

Michael Faraday, an Englishman, and Joseph Henry, an American, made independent discoveries of how this could be done about the same time, but Michael Faraday in 1833 was the first to publish his observations. Although nobody at that time realized the far-reaching importance of this discovery it marked the beginning of an era in which electricity was to become the servant of man.

We can perform similar experiments to those made by Faraday with the simple apparatus shown in Figure 31.1. Plunge a magnet in and out of a coil and observe the deflection of the needle of a galvanometer connected in series with the coil. Vary (a) the strength of the magnet, (b) the number of turns in the coil, and (c) the speed with which the magnet is plunged in and out of the coil. A greater deflection will be found to be produced by a stronger magnet, a larger number of turns in the coil, and a more rapid movement.

We conclude as Faraday did from his observations, that an electromotive force can be induced in a coil by the movement of a magnetic field relative to the coil, and that its strength depends on the rate of cutting of the magnetic lines of force. This process is called *electromagnetic induction*. In fact, when a

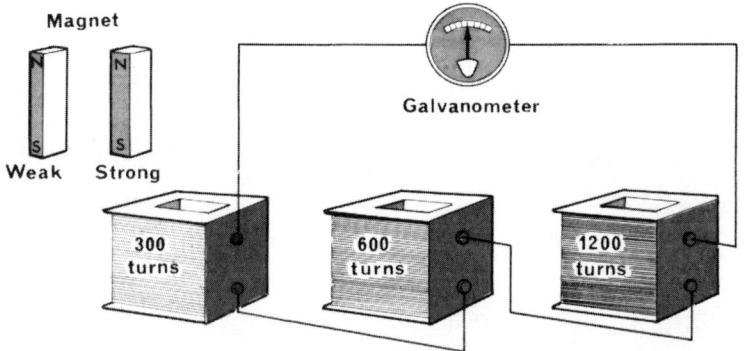

Figure 31. 1. Some apparatus used to demonstrate Faraday's experiments on electromagnetic induction. A weak or a strong magnet can be inserted slowly or quickly into one of the coils. The current produced in the circuit by the induced E.M.F. is indicated by the movement of the galvanometer needle.

conductor cuts magnetic lines of force an e.m.f. is always induced in the conductor. When there is no movement of the magnet or the coil relative to one another no induced e.m.f. is observed. This e.m.f. can cause an electric current to flow if the conductor is part of a closed conducting circuit.

We can also observe when performing these experiments that the *direction* of the induced e.m.f. is dependent on the direction of cutting of the magnetic lines of force by the coil.

The diagrams in Figure 31.2 illustrate the way the magnetic lines of force of a magnet cut by a coil determine the direction of the induced e.m.f. in the coil. The current flows in the coil, owing to the induced e.m.f., in such a direction that the magnetic poles produced at the ends of the coil itself try to stop the relative movement of the magnet and the coil. When inserting the magnet into the coil the movement is against the *repulsion of like poles* and one has to push the magnet into the coil to overcome this repulsion. Does this agree with the principle of the conservation of energy? What would happen if the poles were such as to assist the relative movement of the magnet and coil? On the other hand removing the magnet from the coil induces an e.m.f. in the coil in such a direction that the poles created try to stop the movement; this time one has to pull the magnet out of the coil and work against the *attraction of the unlike poles*. The direction of the induced e.m.f. as explained above was stated by the Russian scientist Lenz in the form of a law, and this is now known as Lenz's law.

The small bicycle generator of electricity in which the coil is rotated in a magnetic field as the wheel turns is an example of the practical use of electromagnetic induction. The e.m.f. induced in the coil, as it rotates, is first in one direction and then in the opposite direction. Thus the induced current flowing through the closed circuit of the small lamp and bicycle framework alternates backwards and forwards. This current is known as an *alternating current* (a.c.), and the number of times it alternates every second is called its frequency.

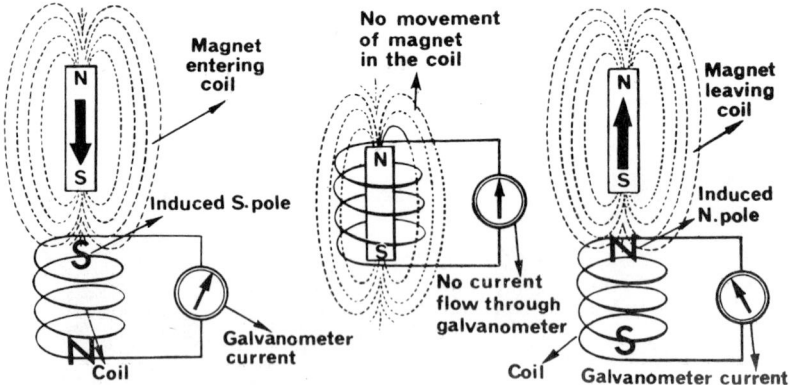

Figure 31.2. The magnetic lines of force due to a magnet are shown cutting a coil as the two move relative to one another. The induced e.m.f. in the coil in both cases is in such a direction that the induced current creates magnetic poles that oppose the movement. The thick arrows show the movement of the magnet towards and away from the coil.

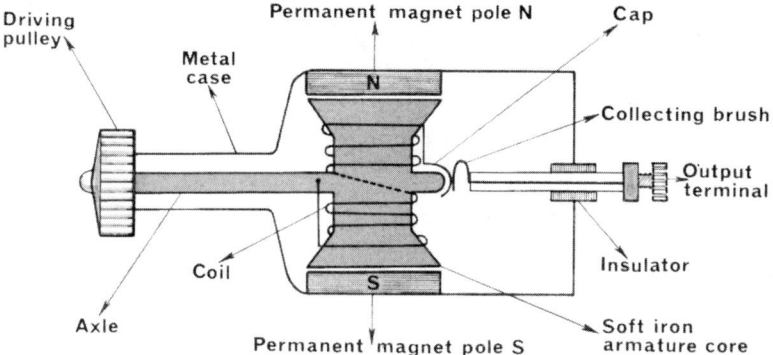

Figure 31.3. A bicycle generator. The magnetic lines of force passing through the coil on the soft iron armature core are completely reversed in direction every half revolution as the coil spins inside the fixed magnet. One end of the coil is connected to the axle and the framework of the generator, and the other end is connected to the small cap and from there to the brush and the insulated terminal.

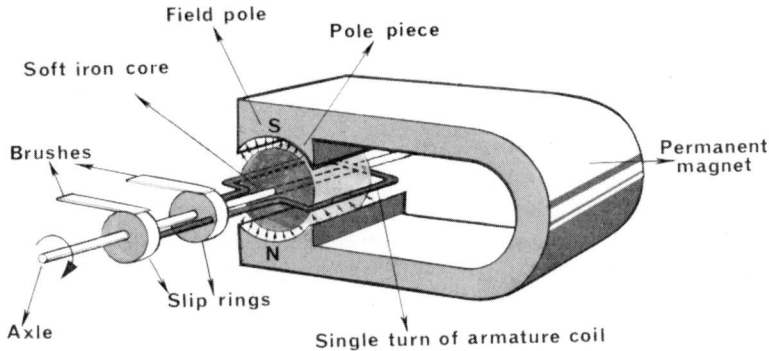

Figure 31.4. A simple generator of alternating current. This diagram shows one turn only of the very many usually built into the coil of the *armature*. Note the curved *pole pieces* and the soft iron *core* of the armature that together make a stronger and more concentrated magnetic field in which the coil can rotate.

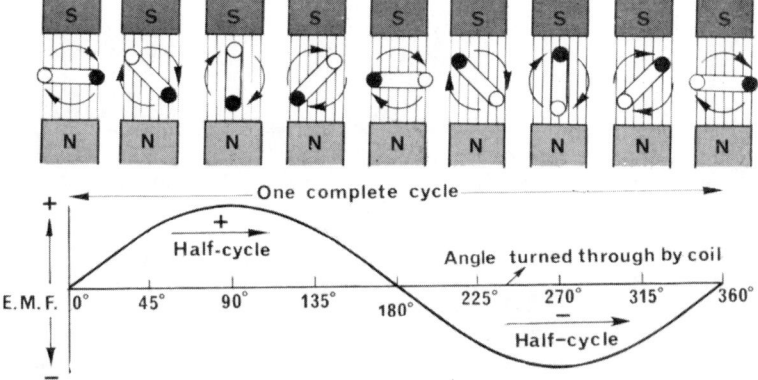

Figure 31.5. Alternating current generator. (Alternator.) The induced e.m.f. at various positions during one cycle—i.e., a complete revolution of the coil in the magnetic field of the fixed magnet.

Let us examine more thoroughly the induced e.m.f. produced in a coil as it spins between the poles of a magnet. When the coil in Figure 31.4 is horizontal and the wire moves along the direction of the magnetic lines of force it cuts none of them and thus no e.m.f. is induced. When the coil is vertical it cuts the maximum number of magnetic lines of force per second and thus the induced e.m.f. is a maximum. See Figure 31.5 on previous page.

Larger generators of alternating electric current or alternators are made with many pairs of magnetic poles and many pairs of coils. The magnetic poles provide the magnetic fields and thus they are usually called *field poles*. They are normally formed by an electromagnet using a *direct current* (d.c.) of electricity. It is difficult to conduct large currents from the spinning coils of the armature so the coils are kept stationary on a *stator*; they are joined together in series, and are connected directly to the output supply. The field poles spin inside the stator on a *rotor* and thus provide a rotating magnetic field. The supply of the direct current electricity to energize the electromagnets of the rotor is provided by another generator called an *exciter*. The exciter is usually mounted on the same axle as the rotor and the d.c. electricity is conducted through slip rings to the electromagnets of the rotor.

D.C. generators or dynamos. The alternators that we have studied can be modified to produce d.c. electricity.

For certain purposes a.c. electricity cannot be used and d.c. is essential. For instance, it is essential to use d.c. to charge a motor-car battery, and then d.c. is drawn from the battery to operate all the electrical equipment of the car. Many electric locomotives require d.c. for their operation as d.c. can be simply and easily controlled. Locomotives using a.c. are now constructed but their speed regulation is more difficult.

Exactly the same armature and magnetic field are used in the dynamo (d.c.) as in the alternator (a.c.). The difference lies in the method of conduct-

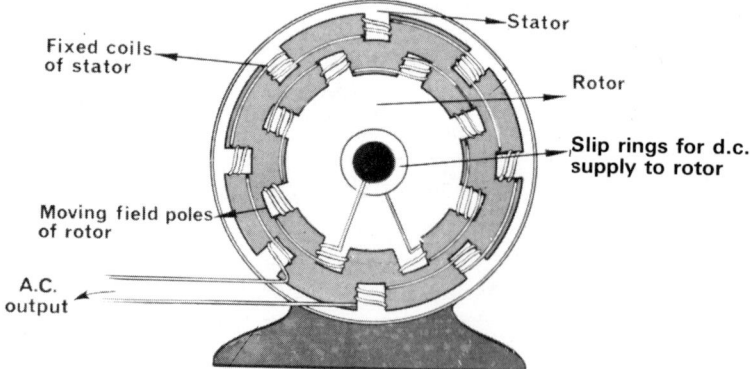

Figure 31.6. An alternator. The stator consists of fixed coils and the rotor of rotating field poles. The slip rings provide the connection between the direct current supply and the rotating electromagnets of the rotor.

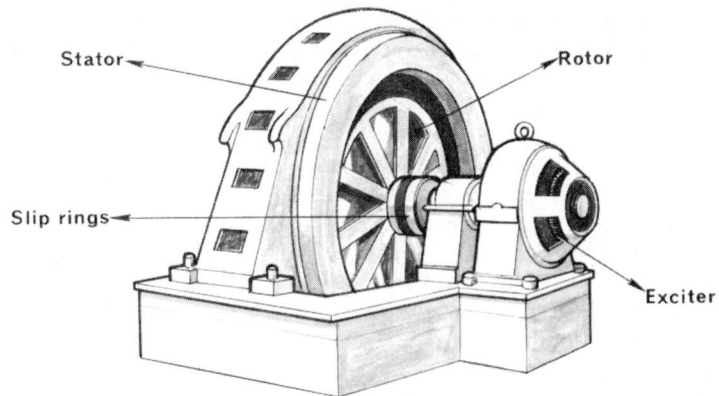

Figure 31.7. A 3 000-volt alternator, The exciter and the slip rings are visible. The main alternating current output is taken from the stator by thick insulated cables. A modern generator is usually completely enclosed so that these parts are invisible.

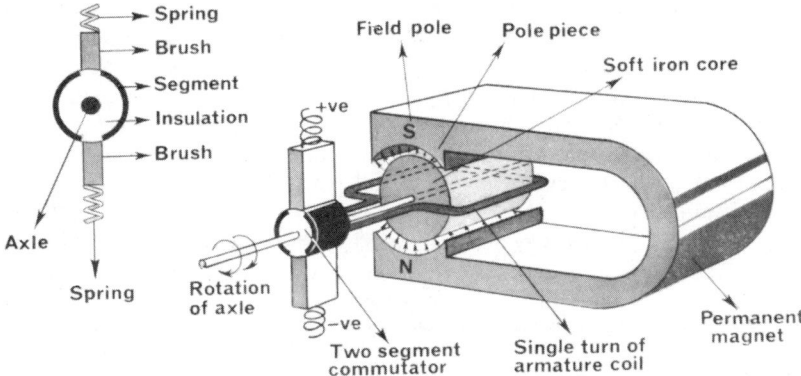

Figure 31.8. A two-segment commutator for d.c. generators shown at the precise moment when the output from the armature coil is zero and the switching over by the brushes from one segment to the following segment is taking place.

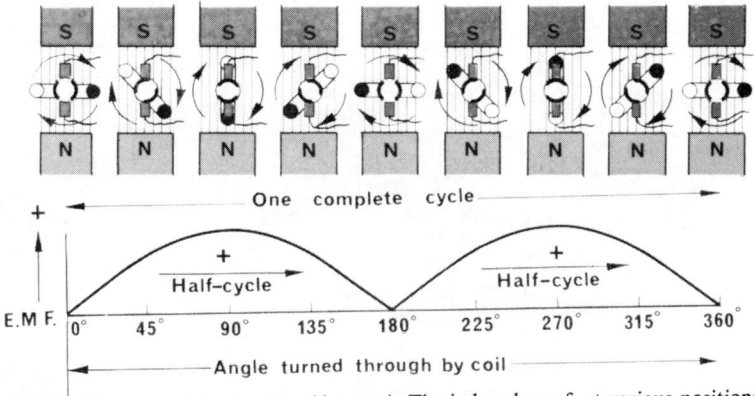

Figure 31.9. Direct current generator (dynamo). The induced e.m.f. at various positions during one cycle—i.e., a complete revolution of the coil in the magnetic field of the fixed magnet.

ing the current out to the external circuit. In the armature coil the current surges back and forth. Therefore to obtain electricity that always flows in the same direction the connections from the spinning coil must be switched over just when the current flow changes direction and is at a minimum. For this switching a split metal cylinder is used which is insulated from the armature axle and turns with it. Brushes are set against opposite segments of this cylinder. The whole arrangement is called a *split ring commutator*. In this way each brush has electricity flowing through it in one direction only. See Figure 31.8.

Some armatures carry many rectangular coils of wire each consisting of a large number of loops or turns of wire. The coils are set at angles to each other so that the wire of some coil is always cutting lines of force. Each coil contributes its share of the total current to the external circuit at regular intervals with the result that a steadier flow of current is delivered. The field magnets of these dynamos are usually excited by means of some of the current supplied by the armature itself. See Figure 31.12.

Electric motors. The general construction and individual components of an electric motor are almost identical with those of a generator. In fact, direct current dynamos and motors are often interchangeable. These machines act as dynamos when mechanical energy is converted into electrical energy, and as motors when electrical energy is converted into mechanical energy. These *dynamotors*, as they are sometimes called, are most convenient because of the double use to which they can be put. Installed on electric trains or trolleys they act as motors when climbing uphill, drawing electricity from the conductor rail. When descending they act as dynamos converting the energy of motion into electrical energy which is converted again, this time into heat energy as the electric current passes through the resistors that form the rheostat. The driver is able to control the speed of the electric train by altering the resistors connected in series with the motor circuit. You can often hear a click as he changes from one resistor to the next when the train gathers speed. In the same way he

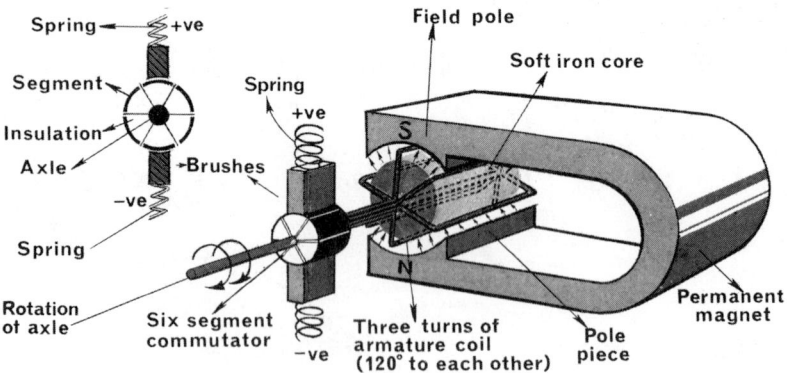

Figure 31.10. A d.c. generator or dynamo having three coils set at 120° to one another. One turn of each coil is shown in the diagram.

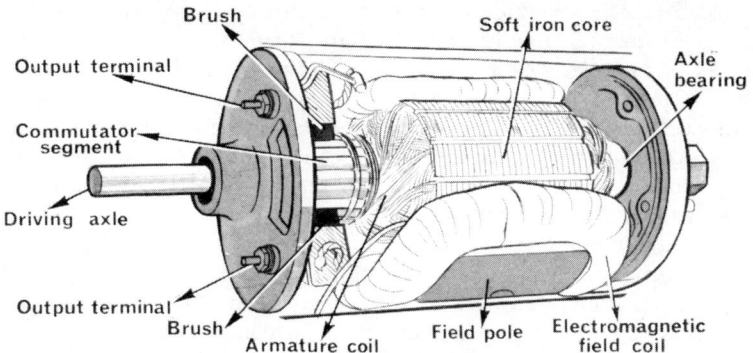

Figure 31.11. A sectional drawing of a motor-car direct current generator. The electromagnetic field coils and field poles, the armature coils and the soft iron core, the commutator segments and brushes, and the output terminals are all shown.

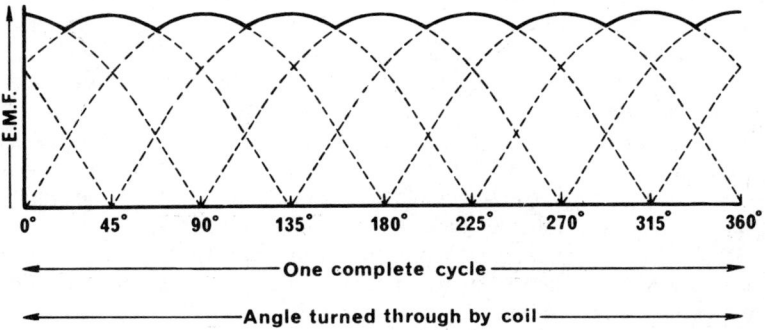

Figure 31.12. The induced e.m.f. of a dynamo having 4 different armature coils connected to an 8-segment commutator. One complete revolution of the armature is shown. The dotted lines indicate the induced e.m.f. of each separate coil and the solid line the total output from the 4 coils together. The greater the number of coils in the armature the smaller the 'ripple' of the total induced e.m.f.

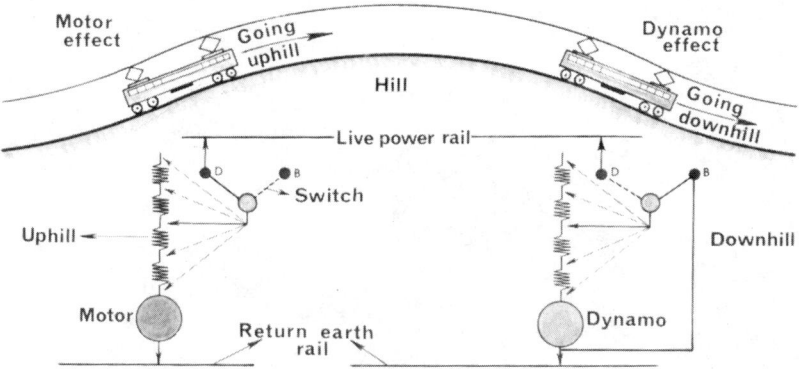

Figure 31.13. The use of dynamotors for electric traction. The switch from 'drive' to 'brake' is shown as well as a few of the possible settings of the rheostat.

can control the braking force when the dynamotors act as dynamos by varying the value of the resistors he connects across them when they are no longer connected to the electric power supply rail. These resistors are often mounted below the floorboards in a train so that they can quickly dissipate their heat.

An electric motor is a practical application of the magnetic effect of an electric current. By noting carefully the direction of the electric current flowing in the armature coil the forces created between the coil and the magnetic poles can be determined. See Figure 31.14.

The d.c. electric motor works as it does because of the attraction and repulsion between the magnetic 'poles' of the armature coil and the field poles. If a d.c. electric motor constructed with electromagnets instead of permanent magnets is supplied with a.c. electricity then the two magnetic poles will attract and repel each other as if d.c. electricity were being supplied. The only difference is that, at a frequency of 50 Hertz, there are two N. poles repelling each other for one hundredth of a second, and then during the next one hundredth of a second there are two S. poles repelling one another. These motors can be called a.c./d.c. motors for they will work with either a.c. or d.c. electricity. They are very useful for small power outputs, such as for vacuum cleaners, fans, and shavers. The larger a.c. motors are not made in this way because they work on an entirely different principle.

A *synchronous a.c. motor* as its name implies keeps its axle rotating so as to remain always in step (or in phase) with the a.c. supply. A simple working model of a synchronous motor is illustrated in Figure 31.16 (*a*) and (*b*). The rotor is a single permanent magnetic needle mounted on a pivot, and the stator is a single electromagnet connected to the a.c. supply. The poles of the stator alternate N S N S N S continuously as the alternating current changes direction. Once the rotor is spun at the correct speed its poles will always be attracted to the unlike poles of the stator as they alternate. In this way if the frequency of the a.c. supply is kept constant the speed of rotation of the rotor will also be constant. Some electric clocks are operated by small synchronous motors and

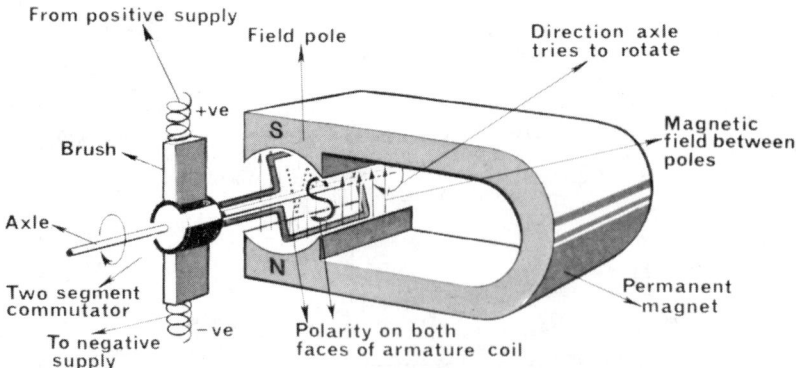

Figure 31.14. The rotation of a one-turn coil of a direct current electric motor. The movement is caused by the attraction and repulsion between the magnetic field of the armature coil and the magnetic field of the poles of the permanent magnet. The polarity of the armature coil is shown by the large dotted capitals N. and S.

their accuracy depends on how constant the frequency of the a.c. supply is kept. The rotor of the synchronous motor illustrated in Figure 31.16 (c) rotates once every fifteen complete cycles of the a.c. supply and therefore a system of gears is used to drive the minute and the hour hands of the clock.

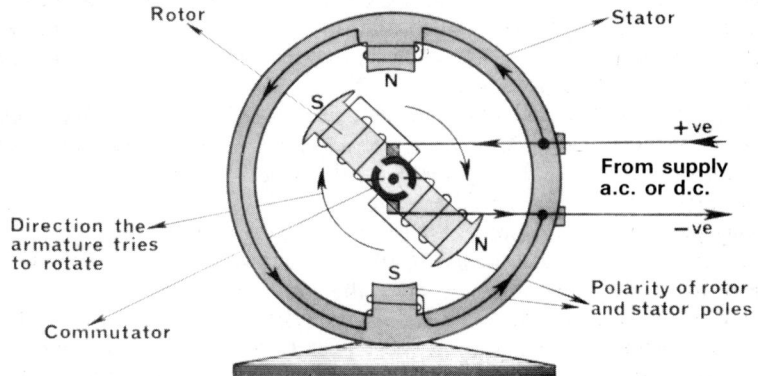

Figure 31.15. An a.c./d.c. electric motor. The electric current energizes the magnetic poles of the stator and the poles of the electromagnet formed by the armature coil of the rotor. The two circuits to the stator and through the commutator to the rotor may be connected in series or in parallel. In either case the two sets of poles attract and repel one another to cause the rotation of the armature whether the current is alternating or direct. The diagram shows the polarity of the stator and the rotor poles at one instant of time.

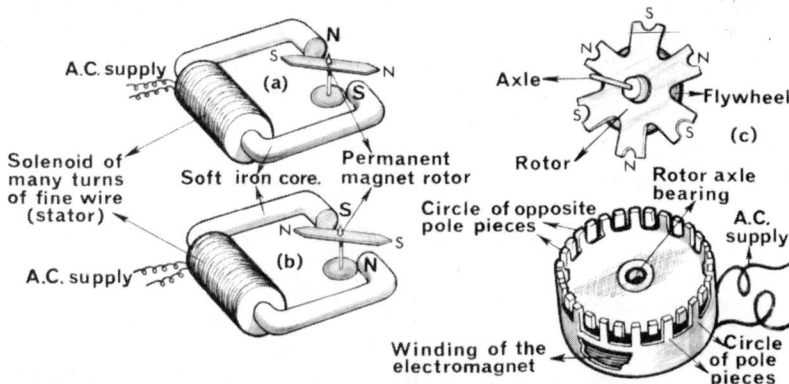

Figure 31.16. (a) and (b) A model of a simple synchronous a.c. motor in the two different phases of the alternating current supply. (c) The synchronous a.c. motor of an electric clock. Note that it has 3 pairs of poles on the rotor and 15 pairs of poles on the stator, and a flywheel to maintain a smooth spinning of the rotor. The solenoid of the electromagnet that forms the stator is wound between the two circles of pole pieces—it can just be seen inside the cutaway sections of the casing.

Three phase alternators. A moving magnetic pole that sweeps along and near to a solenoid mounted on a curved soft iron core produces a rapid change in the direction of the magnetic lines of force passing through the solenoid. This change of the magnetic lines of force induces an e.m.f. in the solenoid. A magnet continually rotating near the curved solenoid produces an induced alternating e.m.f. of the same form as that shown in Figure 31.5. This can be achieved as shown in Figure 31.17 by spinning a magnet about an axle placed at the centre of the curved solenoid.

Three such solenoids mounted on one complete soft iron ring with a rotating magnet in the centre each produce alternating currents that are 'out of step' with one another. In fact, the three induced e.m.f.s formed in the three solenoids may be represented by three waves that are at different *phases*. Hence this arrangement produces *three phase alternating current.*

Large three phase alternators for use in power stations have many sets of three coils (short solenoids) forming the stator, and many electromagnets energized by an exciter forming the rotor. In this way the speed of rotation is reduced so that the rotor can be rotated fairly slowly and is thus prevented from flying to pieces as it might do at high speeds. A three phase alternator is a relatively simple machine to manufacture and three phase motors driven by three phase a.c. are also relatively simple and cheaply made. This is one of the main reasons why three phase a.c. is used almost universally for the supply of electricity on a large scale. Another great advantage of a three phase a.c. supply is that the consumer can draw from it either 240 volts or 415 volts as shown in Figure 31.19. In order to balance the outputs from the three phases of the generator an effort is made to draw equal currents from each of them. In order to do this houses in the same district are connected in rotation to the three phases. Powerful electric motors are wired to all the three phases.

Three phase a.c. electric motor. A rotating magnetic field can be formed by connecting three coils one to each phase of a three phase a.c. supply. If these

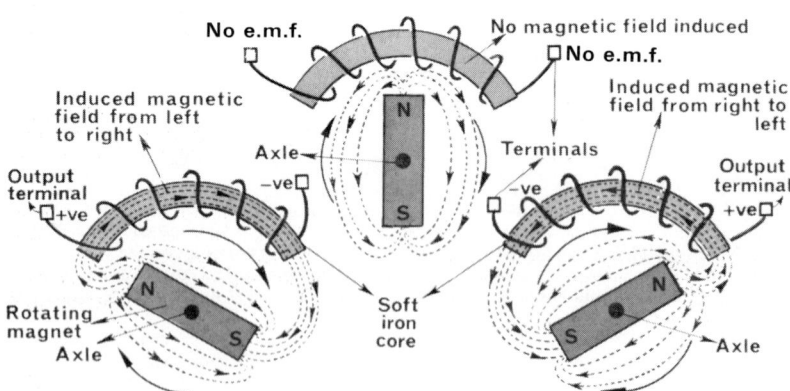

Figure 31.17. How one alternation of an alternating current can be produced by rotating a magnet near a curved solenoid. The directions of the magnetic lines of force are marked by dotted lines.

coils are separated 120° apart round a circle the rotating magnetic field they create will draw round with it a freely suspended permanent magnet or, by induction, a cage of stout copper rods. This is the principle on which induction motors are made, and they are self-starting. See Figure 31.20. The induction motor may be considered to be a transformer in which the secondary coil is mounted on a shaft so that it can rotate. When the current flows in the primary coils, called the stator, an induced current flows in the secondary coils, called the rotor. The magnetic field produced by the primary coil reacts with the magnetic field of the secondary and produces the rotation of the rotor. The term 'induction' indicates that there is no electrical connection between the

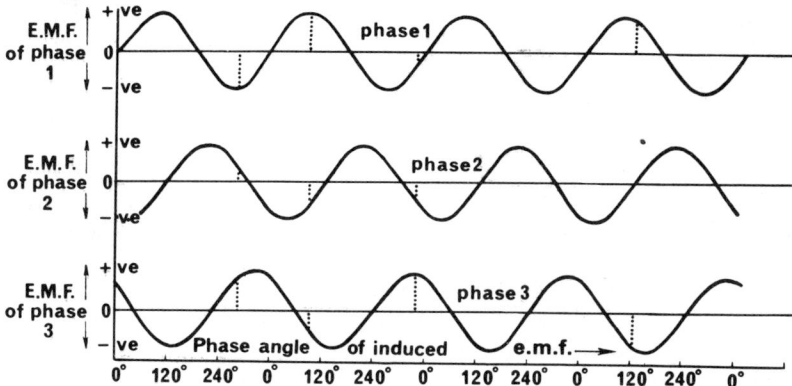

Figure 31.18. The wave forms of three separate induced e.m.f.s each out of phase by one-third of a cycle or 120°. Wherever the total e.m.f. of the three phases is measured (shown by the dotted lines) it is zero, because there is as much in the positive direction as there is in the negative direction. Four cycles are shown in the diagram.

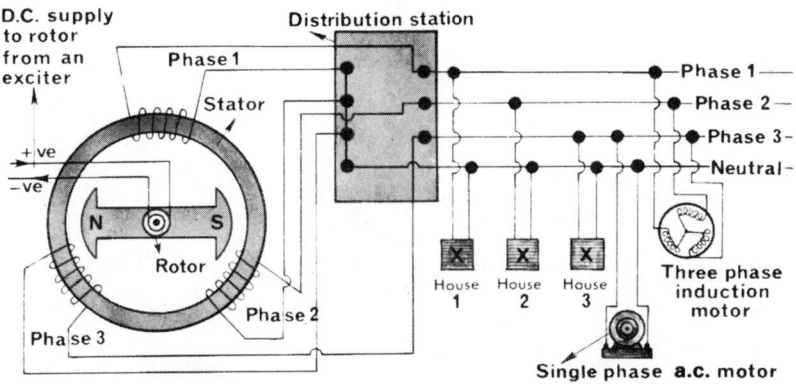

Figure 31.19. A diagram of a three-phase alternator supply system and the distribution of the electrical power generated. The fourth wire called *neutral* is connected to each coil of the alternator and to each house. Each house is also connected to one of the phase wires and the three phase motors to each of the three phases. Between one phase and the neutral on the British system the maximum e.m.f. is 240 volts, and between any one phase wire and another the maximum e.m.f. is 415 volts.

stationary and the rotating parts of the motor, and that the two parts are coupled only by induction.

This induction type of three phase electric motor is probably the most frequently used one for driving medium and heavy machinery in industry.

Transformers. When the transmission of electrical power has to take place over long distances of several miles the energy loss in the form of heat along the wires is so large that we transmit the power as a small current at a very high voltage in order to minimize these losses. The conversion from the alternating *low tension* (low voltage) supply of the alternators to a *high tension* (high voltage) supply is accomplished by the use of transformers. Direct current electricity cannot be easily transformed. This is the strongest reason for using a.c. electricity rather than d.c. electricity for long distance transmission supply.

The transformer is another example of applied electromagnetic induction. An alternating e.m.f. connected to the *primary* coil of a transformer causes an alternating current to flow through it; this in turn causes an alternating magnetic field in the iron core, an alternating e.m.f. in its *secondary* coil, and finally an alternating current in its secondary circuit if the circuit is a closed one.

The change in voltage depends on the relative number of turns of wire in the two coils. The ratio of the voltage of the supply connected to the primary coil to the voltage induced in the secondary coil is, apart from small losses, the same as the ratio of the number of turns of wire in their respective coils.

$$\frac{\text{Primary e.m.f.}}{\text{Secondary e.m.f.}} = \frac{\text{Number of Primary turns}}{\text{Number of Secondary turns}}$$

Modern transformers are so well made that there is little loss due to stray alternating magnetic field outside the iron core and little difficulty in alternating rapidly the direction of flow of the magnetic lines of force in the iron core. If you examine any transformer you will find that the soft iron core is

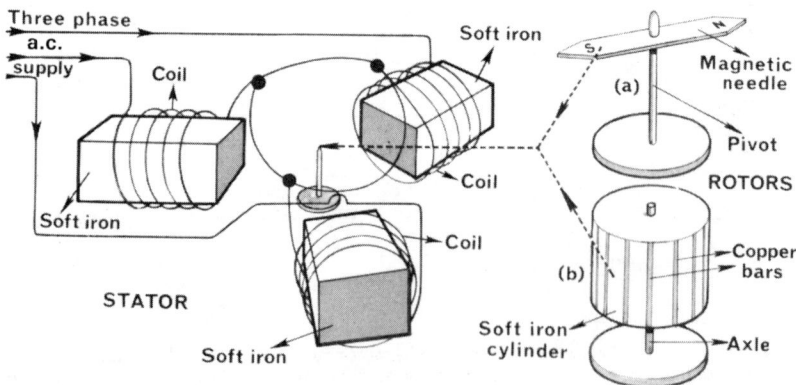

Figure 31.20. A working model built to illustrate a three phase a.c. electric motor. The rotor in (a) is a permanent magnetic needle, and in (b) is a 'squirrel-cage' of copper bars mounted on a vertical axle around a cylinder of soft iron. The stator consists of three electromagnets arranged symmetrically round a circle. One terminal of each electromagnet is connected to one of the three phases of the supply. The other three terminals are joined together.

made of a number of thin insulated laminations of soft iron clamped tightly together. This reduces very considerably the loss of energy due to heating in the iron core. Thus there is hardly any loss of energy in the transformer; the electrical energy supplied to the primary coil is almost all given out again in the secondary coil. If we assume no loss at all, then the energy input to the primary coil equals the energy output from the secondary coil. If we remember that Energy = Voltage × Current × Time, then we can form the following equations:

ENERGY INPUT TO PRIMARY COIL = ENERGY OUTPUT FROM SECONDARY COIL

or:

VOLTAGE × CURRENT × TIME = VOLTAGE × CURRENT × TIME
in PRIMARY COIL   in SECONDARY COIL

or considering the electrical POWER (energy per unit time)

POWER INPUT   =   POWER OUTPUT

or:

VOLTAGE × CURRENT = VOLTAGE × CURRENT
in PRIMARY COIL   in SECONDARY COIL

When transmitting electrical power the large alternating current at low tension (voltage) from the alternators is converted by a 'step-up' transformer to a small current at high tension (voltage) with very little loss of power if the transformer is efficient. The heat loss along the overhead wires is proportional to the square of the current flowing (see Chapter 29), and this is why the heat loss due to a small current, even at a high tension, is so much less than the heat loss due to a large current. For this reason electrical power is transmitted over long distances at high voltages and low currents.

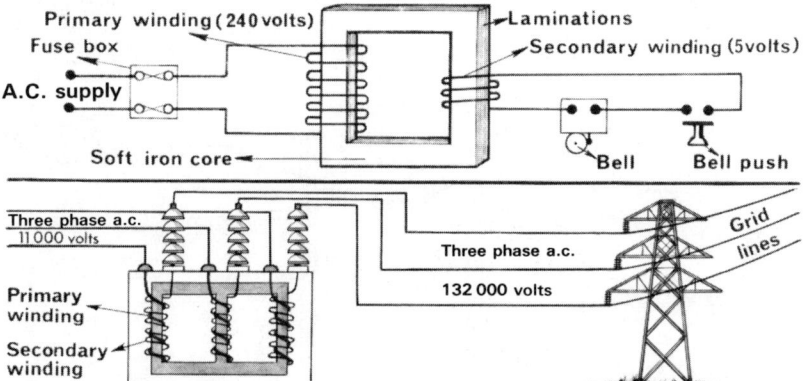

Figure 31.21. A bell transformer connected in its circuit. This is a 'step-down' transformer because it steps down the voltage from 240 to 5 volts. It has more turns in the primary coil or winding than in the secondary coil or winding. A 'step-up' transformer has more turns in the secondary coil than in the primary coil.

Figure 31.22. A three phase a.c. transformer. This particular transformer 'steps up' the voltage from 11 000 to 132 000 volts for transmission on the overhead grid system.

Distribution of three phase a.c. takes place at high tensions over a network of wires called a *grid*. Three phase transformers raise the tension to 132 000 volts for the grid, to 275 000 volts or 400 000 volts for the *supergrid*. The electricity at these high tensions is carried by wires suspended by long insulators from pylons. At suitably placed sub-stations, where there are many consumers of electricity, stepdown transformers convert the high tension electricity to low tension electricity for domestic use.

The induction coil. This is a transformer so designed that it can work when connected to a source of d.c. electricity. It consists of a primary coil of very few turns of thick wire wound on a soft iron core and a secondary coil of many turns of fine wire wound around the primary coil. There is a mechanical 'make and break' in the primary circuit that works like the vibrator of an electric bell.

When the circuit is closed at 'make' the current grows in the primary coil, inducing an e.m.f. in one direction in the secondary coil. This growth of current takes a small but a definite time to take place. At 'break' the current ceases to flow very very rapidly, much more rapidly than it grows at 'make'. One reason for this rapid decay of current is that a capacitor is inserted across the contact points to eliminate the spark that would otherwise form at 'break'. In this way the induced e.m.f. at 'break' is very much greater than it is at 'make', and for all practical purposes the induced secondary e.m.f. is unidirectional and only occurs at 'break'. The whole cycle of 'make', 'on', 'break', and 'off' takes place quickly—perhaps fifty times a second.

Induction coils have many applications. They are used to supply the sparks for the engine of a motor-car, the high voltage power for X-ray tubes, and to assist in atomic research and in medicine.

The ignition coil of the motor-car. Advantage is taken of the large e.m.f. induced at 'break' in an induction coil to produce a spark just at the right moment to fire the explosive mixture in the cylinder of a motor-car engine. A

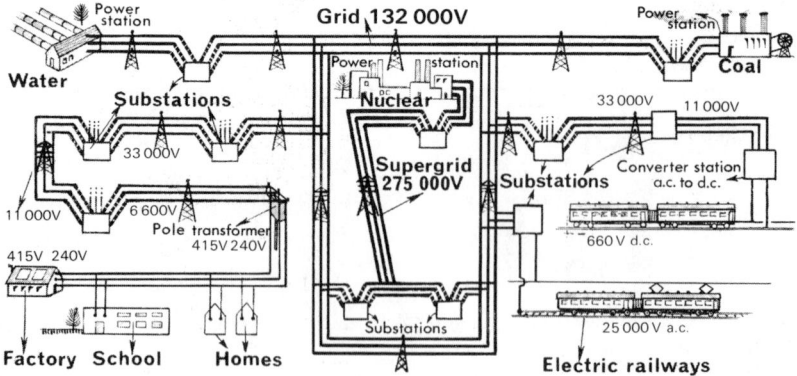

Figure 31.23. The generation, transmission, and distribution of electricity by a 'grid' system.

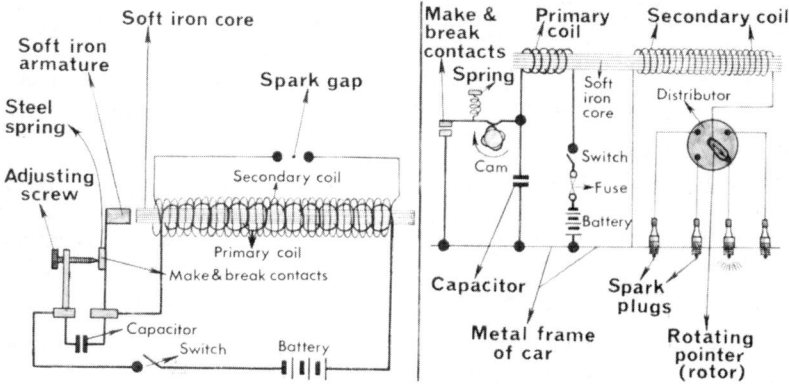

Figure 31.24. The essential parts of an induction coil.

Figure 31.25. How an induction coil is used in the ignition system of a motor-car. The primary and secondary coils with the soft iron core form the ignition coil in this case. The 'make and break' mechanism, the capacitor, and the distributor of the high tension induced e.m.f. form another unit of the system.

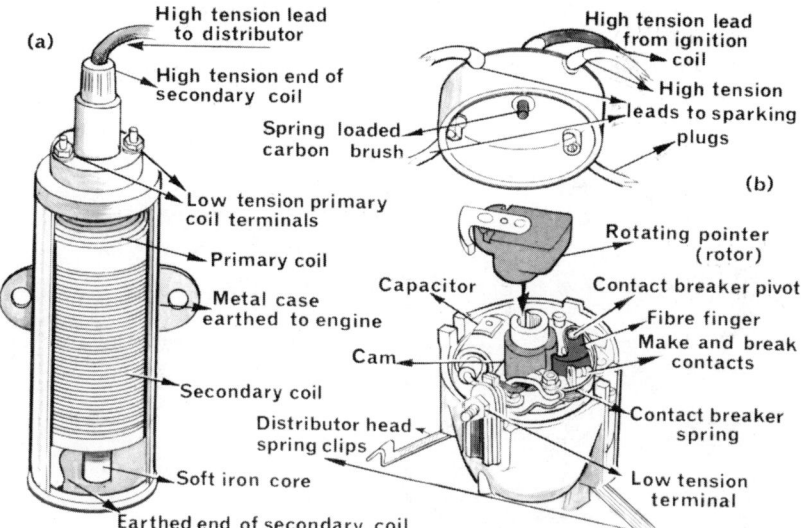

Figure 31.26. The units of the ignition system of a motor-car. (*a*) The ignition coil. (*b*) The contact breaker, capacitor, and distributor. The rotor as it turns almost touches, in the correct order, the terminals in the distributor head connected to the sparking plugs.

spark requires only a very small current but a voltage of perhaps 15 000 volts. The precise moment of firing is determinded by a cam, connected directly to the main axle of the engine. The cam opens a pair of contact points and thus breaks the primary circuit. As the engine turns, the sparking plugs, one after the other, are connected to the secondary coil of the ignition coil. The rotating metal pointer of the distributor connects the lead to the correct sparking plug.

The electric cattle fence is another common application of the induction coil. The 'make and break' in this case is made by contacts attached to a spring controlled oscillating flywheel. The high tension induced e.m.f. in the secondary coil, connected directly to the bare wire 'fence' slung on insulators, could give a strong 'shock' to any animal standing on the earth and completing the secondary circuit by touching the wire.

A rectifier is a one-way conductor. It offers a low resistance to an electric current flowing in one direction and a high resistance to one in the opposite direction. It is often used to convert the a.c. of the mains supply into d.c. There are many reasons why d.c. is needed; for example, the charging of an accumulator, electroplating, and the working of small reversible electric motors.

One type of rectifier is the *rotary-convertor* which consists of an a.c. motor, the armature of which is coupled with, or combined on the same axle as, the armature of a d.c. dynamo. The a.c. motor drives the dynamo which is designed to produce the d.c. voltage required.

Another type is the *metal rectifier* which is made of discs of two substances pressed together. There are several pairs of substances which when placed together in series in an a.c. circuit rectify the a.c. to produce d.c. Arranged in this way *half-wave rectification* is produced. By connecting two or four of these metal rectifiers in other ways *full-wave rectification* is obtained.

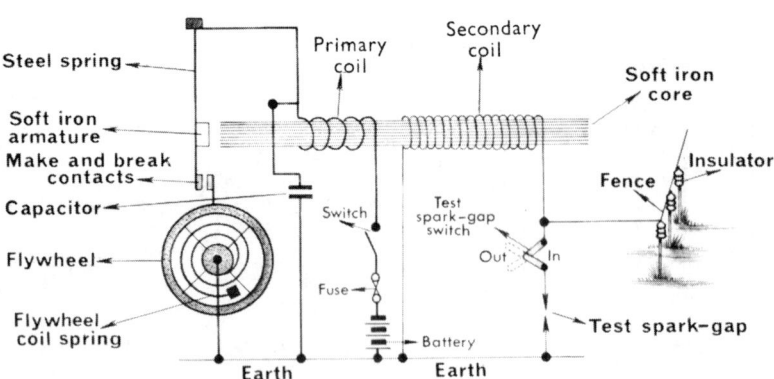

Figure 31.27. How an induction coil is adapted to provide the high tension e.m.f. needed to give a strong 'shock' when a electric cattle fence is touched.

A diode (see Figure 26.14) allows the passage of electrons one way only and thus an electric current in the opposite direction only. These are used extensively in radios and television receivers for the production of electric power.

Crystals of various substances have been found to be able to rectify an a.c. In the next chapter a *germanium* crystal is described rectifying the high frequency a.c. of a radio oscillatory circuit. It is called a germanium diode and acts as a *detector* (see Figure 32.5). A *transistor* also can be made to rectify and do other things at the same time. (See Figure 32.8).

The telephone. Alexander Graham Bell devised the first practical telephone in 1876 but few people at that time realized how important it was to become.

Essentially a telephone system needs three parts: (1) a *microphone* to enable the sound waves of the speaker to produce an electric e.m.f. or current that varies in the same way as these sound waves, (2) an *electric circuit* connecting the microphone and the listener's apparatus, and (3) an *earpiece* or receiver to convert the varying electric current back again into sound waves.

The telephone *microphone* has a thin diaphragm that vibrates as sound waves strike it and behind it there is a small box filled with hard carbon granules that form part of an electric circuit. The variations of the electric current in this circuit correspond with the vibrations of the diaphragm and the carbon granules. This carbon type of microphone used in telephone receivers has a poor response, making the human voice sound 'mechanical', and is therefore unsuitable for quality recording.

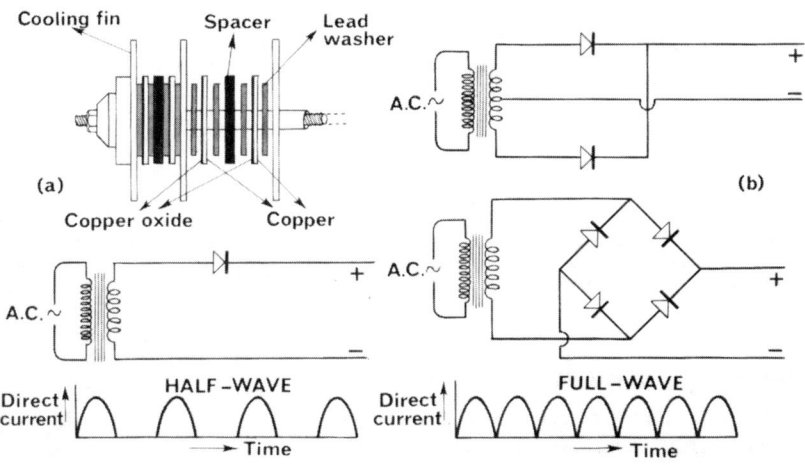

Figure 31.28. Metal rectifiers. (a) A series of copper oxide rectifiers with one pair opened out to show clearly the lead washers, metal spacers, and cooling fins. The current flows from copper oxide to copper metal as shown in the diagram below producing half-wave rectification. (b) Two rectifiers connected to the ends and the centre tap of the secondary coil of a transformer, and below a bridge-circuit of four rectifiers. Can you understand how both these two circuits give full-wave rectification?

The varying electric current from the microphone flows through the primary coil of a step-up transformer where it sets up in the secondary coil an alternating e.m.f. that has the characteristics of the sound waves that enter the microphone. The alternating current thus produced passes along the line wires to the distant receiver.

The *earpiece* consists of a small permanent magnet which attracts and very nearly touches a thin flexible steel diaphragm. There is a coil of many turns of fine wire wound on each pole of the magnet and through these coils the alternating electric current from the transmitting station passes. This varying a.c. varies the strength of the magnetic field of the magnet and this causes the diaphragm to vibrate and emit sound waves that correspond to the original sound waves striking the microphone.

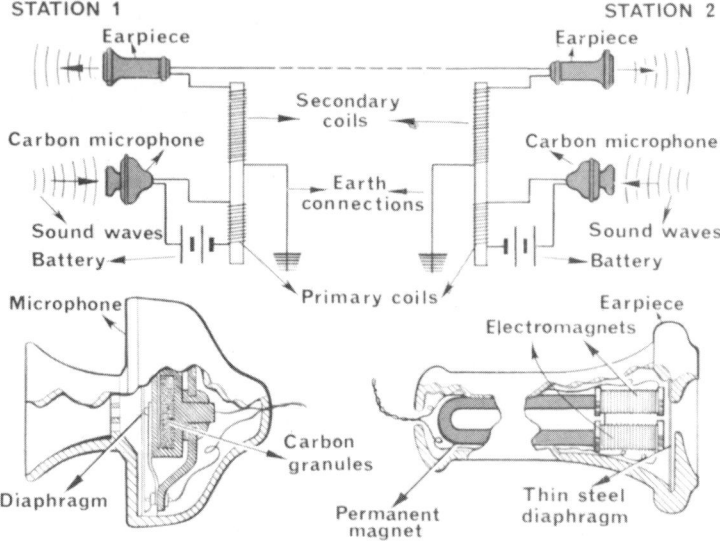

Figure 31.29. A simple two-way telephone circuit. Sometimes one wire only is used between the stations as the circuit can be completed through good earth connections. Above are simplified diagrams of a microphone and an earpiece showing the electrical circuits in each instrument.

Questions on chapter 31

1. What are the essential conditions for inducing an e.m.f. in a conductor? What factors determine the strength of the induced e.m.f.? How can the induced e.m.f. produce an induced current, and what determines the magnitude of the induced current so formed?

2. How is it possible to predict, without the use of any apparatus, the direction that an induced current would flow in a conductor?

3. Draw and name the essential parts of (a) an a.c. generator, and (b) a d.c. generator.

4. What kind of induced e.m.f. is always produced in the wires of the armature of any generator?

5. Explain, with the aid of diagrams, how the induced current produced is conducted away from the armature in the case of (a) an a.c. generator, and (b) a d.c. dynamo.

6. Why do commercial generators of very large currents have many magnetic poles?

7. In what directions can a wire be moved in a magnetic field without producing any induced e.m.f.?

8. An electric motor is mechanically coupled directly to an electric generator. Is it possible to drive the electric motor with the electric current produced by the generator? State clearly the reasons for your answer.

9. Draw a labelled diagram of a transformer, such as is used in public electricity supply, which is capable of dealing with high voltage and high power.

Give one example of the use of (a) a step-up transformer, and (b) a step-down transformer.

10. An electric light bulb is rated at 6 V, 0.3 A. This is operated by means of a suitable transformer from 240 V a.c. mains supply. What is the turns-ratio of this transformer? What power does this lamp consume, and how long will it take to use 1 unit of electricity? (Assume there are no losses in the transformer.)

11. Draw a circuit diagram of a low voltage, mains-operated, electric bell system. Mark in clearly the position of the bell-push. (Details of the bell itself are not required. It should be shown only as a rectangle with two terminals.)

12. Why is the core of a transformer made of soft iron and not steel? Why is the core made of thin laminations and not of one solid piece of soft iron?

13. Draw a diagram of an oil-cooled transformer used for dealing with high-voltage supplies of electricity.

Show diagrammatically the distribution of electricity from a generating station to the national grid and thence to a house.

14. An alternator transmits 20 amperes at 550 volts along a cable to a supply point. The same power is transmitted at 11 000 volts along the same length of a different cable. Give reasons why the cost of the cable in the second case is cheaper than in the first case.

A house is supplied by 240 V a.c. mains. State (a) the type, and (b) the turns-ratio of a transformer which is required to run a 12 V model electric train from the mains supply.

15. Draw a diagram of an induction coil and make clear the essential differences between it and a transformer.

16. What purpose is served by the condenser connected in parallel across the spark gap of an induction coil?

17. Explain how it is possible for you to carry on a conversation with another person using only two telephone receivers and no battery. You are several hundred yards away from the other person and you are connected by two thick electrical wires. Draw the electrical circuit that would be required. Show how the transmission could be improved if each of you were provided with a transformer of suitable design. How are these transformers connected in the circuit?

32. Electronics

This subject deals with the practical applications of those devices that involve the movements of electrons. Thermionic valves, transistors, cathode ray tubes, and the photo-electric cell are amongst the most important of these applications which affect the lives of millions of people.

Radio communication is the transmission of sound by means of electromagnetic waves. Radio has been made possible by the efforts of many workers from different lands freely exchanging their knowledge.

In chapter 24 we read that as a result of the work of the Scottish physicist James Clerk Maxwell and the German Heinrich Hertz, the Italian Guglielmo Marconi designed the first apparatus that was able to send messages through space on a commercial scale. It was on 12 December 1901 that Marconi transmitted signals from Poldhu in Cornwall, England, to St. John's in Newfoundland. The Englishman John Fleming invented the thermionic valve which was later much improved by the work of an American, Lee De Forest, who introduced the grid into Fleming's diode and turned it into a triode. The diode and the triode were the great inventions that made it possible to develop radio as a practical proposition. Radio and television that followed now play an important part in shaping our civilization.

Radio waves like heat and light waves are electromagnetic waves or pulsations of energy that travel at 3×10^8 m s^{-1}. The only difference between radio waves and heat and light waves is in the frequencies of the waves.

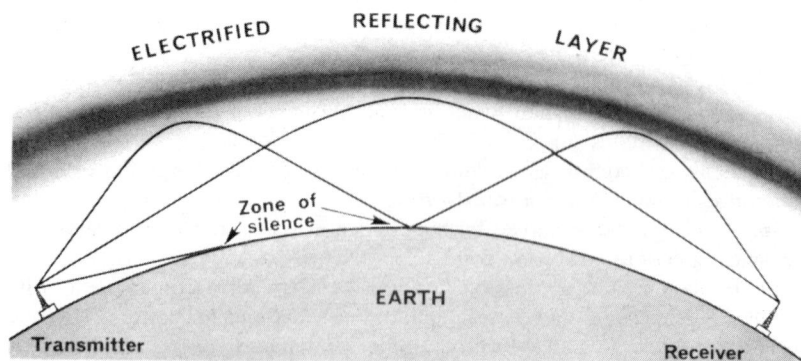

Figure 32.1. How ground waves and waves reflected from the electrified layers travel outwards from a radio transmitter.

The following relationship holds for radio waves as for all other electro-magnetic waves:

Velocity of the waves $(3 \times 10^8 \text{ m s}^{-1})$ = Frequency × Wavelength.

Radio waves are given different names according to their different wave-lengths and frequencies. It is the usual practice to state wavelengths in metres except that microwaves are often stated in centimetres.

Wave group		Wavelength in metres		Frequency in kilohertz	
Long	LW	10 000–1 000		30–	300
Medium	MW	1 000–	100	300–	3 000
Short	SW	100–	10	3 000–	30 000
Very high frequency	VHF	10–	1	30 000–	300 000
Microwaves	UHF	1–	0.01	300 000–30 000 000	

As you can see from this table there is a very wide range of radio waves at present in use, the longest wavelength being a million times longer than the shortest.

Radio waves travel in straight lines, but the long, medium, and short waves are reflected by natural electrified layers high above the earth's surface. These reflected radio waves can travel a long way round the earth. The electrified layers vary somewhat as they are formed by the rays from the sun and there-fore reception varies between day and night and between winter and summer. Both radio ground waves and waves that are reflected can nevertheless leave some 'zones of silence' where reception of the waves is impossible.

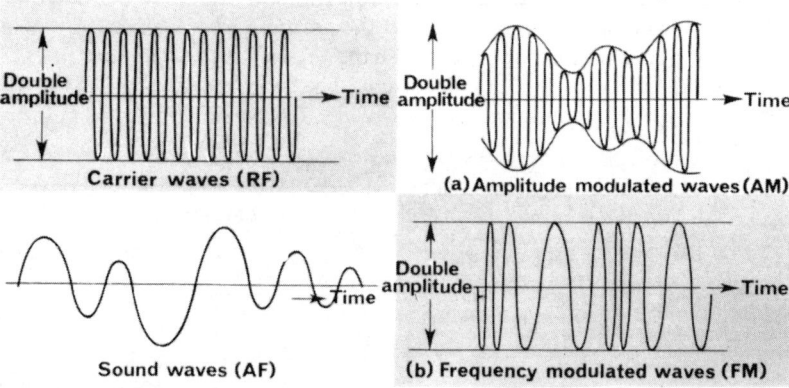

Figure 32.2. These diagrams of amplitude and time show the result of modulating a particular carrier wave by sound waves to produce (*a*) an amplitude modulated (AM) wave, and (*b*) a frequency modulated (FM) wave. The RF carrier waves have a frequency of many kilocycles per second and for the sake of clarity these have been opened out con-siderably along the time axis. The AF sound waves modulate the amplitude of the RF carrier waves to produce the envelope shown in (*a*), and modulate the frequency of the RF carrier waves to produce the constant amplitude and variable frequency waves shown in (*b*).

Look again at the frequencies of the radio waves—they are all very much higher frequencies than those of the sound waves that we can hear. We say that radio waves are propagated at *radio frequencies* (RF) and these are *supersonic*, that is, above 20 kHz the maximum frequency of any sound that the human ear can detect. We also refer to these RF waves as being the *carrier* waves of radio telephony and telegraphy, because these waves carry the transmission. We alter the shape of (or *modulate*) the carrier waves with the sound waves that we can hear at *audio frequencies* (AF).

Modulation of the RF carrier waves in sympathy with the audio frequencies can be done in two ways—(1) by varying the *amplitude* (AM), that is, the maximum potential reached by the aerial at the crest of each wave, and (2) by varying the *frequency* (FM).

Frequency modulation of carrier waves is used for high quality sound broadcasting because it is not upset by atmospherics and other interference. The BBC use frequency modulation for their VHF and UHF sound broadcasts.

Radio transmission and reception can be shown simply in the form of a block diagram, see Figure 32.3. The form of the waves is shown below each section. (1) The girl speaks into the microphone and the sound waves produce the AF electrical wave pattern. (2) The RF oscillator produces the RF carrier waves. (3) The AF waves modulate the RF carrier waves. (4) The resulting AM waves are amplified and passed to the transmitting aerial and earth system. (5) The AM waves are propagated through space getting weaker and weaker the further they travel from the transmitter and towards the receiver. (6) The weak AM waves are collected on the receiving aerial and earth system. (7) The RF carrier waves enclosed in their AF envelope forming the AM waves are amplified. (8) The AM waves at this stage are relatively strong and are converted by the detector into a direct current of electricity pulsating in tune with the shape of the AF modulation. (9) This pulsating direct current is amplified. (10) The loudspeaker responds to the shape of the AF modulation and not to the RF carrier waves because the

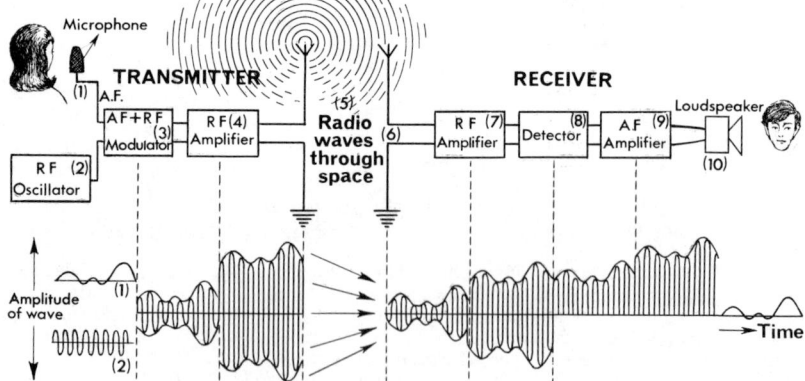

Figure 32.3. A block diagram showing the units used to transmit and receive radio waves. The waves outlined below and between each unit indicate the shape of the waves transmitted from one unit to the next in the whole process of radio communication.

latter are too fast to disturb the cone of the loudspeaker. The AF modulation is now the pattern of the sound waves emitted by the loudspeaker and heard by the boy.

Although there may appear to be many different operations involved in this 'simple' process of transmitting a radio telephone message most modern equipment has many more units. Thermionic valves or transistors are used in the units to amplify, detect, and modulate both AF and RF waves in their appropriate positions. Much work has been done by radio engineers to ensure that the shape of the sound waves entering the transmitter is the same as the shape of the sound waves leaving the receiver. We all know how well they have succeeded in this.

The wiring diagram of a radio receiver and its units may look very complicated but there are only four essential components apart from some amplifying circuits. They are (1) the *aerial–earth* circuit that collects the incoming waves, (2) the *tuned oscillatory circuit*, coil and capacitor, that selects the particular station that one wishes to receive, (3) the *detector* that changes the high frequency alternating current into a pulsating direct current, and (4) the *speaker* that changes the pulsating direct current into sound waves.

A coil of wire connected to a capacitor forms an *oscillatory circuit* for electromagnetic waves. It is rather like a child's swing or the balance wheel of a watch in that it will oscillate back and forth at a fixed frequency. The only way of changing its frequency is by changing either the size and number of turns in the coil or the size of the capacitor.

The oscillatory circuit in a receiver is 'tuned' by a variable capacitor (one whose size can be varied by the simple process of turning a knob on which a scale is often marked) until the circuit oscillates continuously in sympathy with

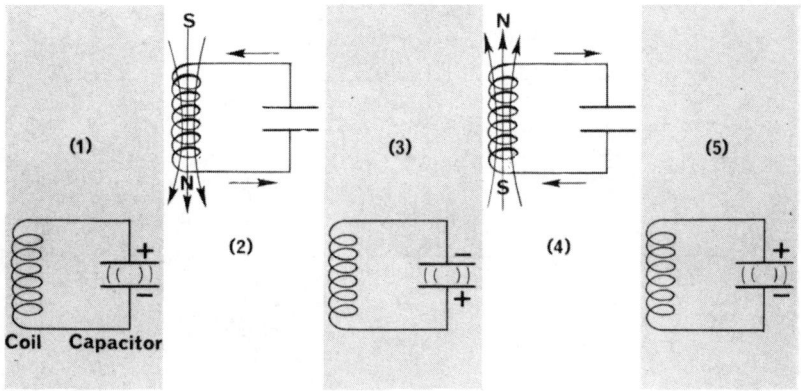

Figure 32.4. An oscillatory circuit. (*1*) The plates of the capacitor are charged and no current flows through the coil of wire. There is no magnetic field in the coil of wire. All the energy is electrical and is stored on the plates of the capacitor. (*2*)The current flows through the coil and forms a magnetic field there. The energy is now all magnetic as no charge remains on the plates of the capacitor. (*3*) and (*4*) The process is repeated in the reverse direction. (*5*) The circuit then returns to stage (*1*). The nature of the energy of the oscillatory circuit alternates; first it is electric, and then magnetic, then electric, and so on indefinitely.

R

the frequency of the incoming radio waves. The small electrical impulses collected by the aerial and earth system keep up these oscillations and do not allow them to fade away.

The alternating potentials from the oscillatory circuit are then fed to the detector which allows the current to pass through it in one direction only. The detector may be a crystal of germanium, or a diode of the type discussed in chapter 26. Thus pulsating direct current eventually arrives at the loudspeaker where it generates sound waves.

The triode thermionic valve, like the diode, allows the passage of electrons in one direction only so that it can 'detect' in the same way as the diode or the germanium crystal. But the triode has a *grid* inserted in the path of the electrons and this controls the electron stream flowing from the cathode to the anode. A negatively charged grid reduces the strength of the electron stream and a positively charged grid increases the strength. The addition of the grid enables the triode to act as an amplifier to the electron stream and causes a stronger pulsating direct current to pass through the loudspeaker. Figure 32.6. A triode can therefore be used as both a detector and an amplifier at the same time in a radio receiver. A simple radio receiver circuit using a triode is shown in Figure 32.7. The same oscillatory circuit of coil and variable capacitor, as used in the germanium diode receiver shown in Figure 32.5, supplies its alternating potentials to the grid through a fixed capacitor. Any electrons that happen to hit the grid by chance can leak away through the high resistor of a million ohms resistance. The plate circuit includes the speaker and a battery to supply the potential difference between anode and cathode.

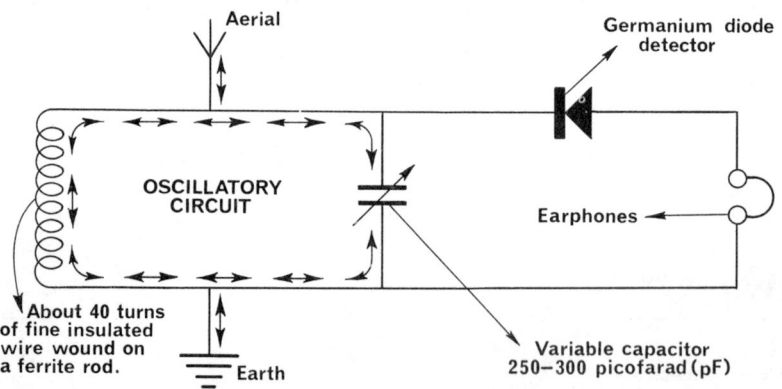

Figure 32.5. A simple radio receiver using a germanium diode as the detector. The small arrows show the directions of the oscillatory currents produced by the aerial-earth system. The values marked of the components in the oscillatory circuit are approximate and are intended to enable the reader to construct a receiver for amplitude modulated medium wave broadcasts. The actual number of turns on the coil will be determined by the wavelength of the particular broadcasting station to which you desire to tune. A strong local broadcasting station can be heard faintly if the size of the capacitor is correctly adjusted between 250 and 300 picofarads. The reception will be weak because there is no other source of energy but the incoming signal to operate the earphones. However if the output is led to an audio amplifier of any kind (eg a tape recorder) excellent reception can be heard and even recorded.

A three electrode *transistor* can replace the triode thermionic valve in a radio receiver. There is little difference in the actual circuit required. The transistor has the advantage that it has no filament to heat and therefore needs no low tension filament battery, nor does it require a very high voltage battery to supply the power to make it work. The transistor is also almost indestructible and shock resistant. It is extremely small and this makes it very serviceable for use in small portable receivers. See Figure 32.8.

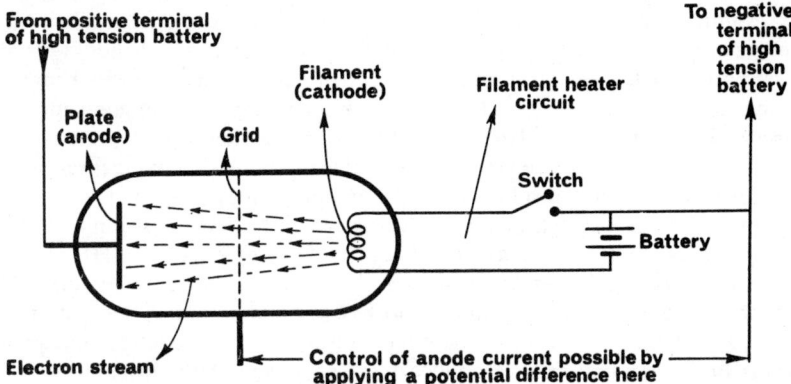

Figure 32.6. A triode thermionic valve. The grid, being close to the cathode and in the path of the electron stream to the anode, exerts a strong control on the passage of electrons. When it has a negative charge it can stop the electron stream, and when it has a positive charge it can draw a large number of electrons out of the cathode and accelerate them towards the anode. Compare this triode with the diode shown in Figure 26.16.

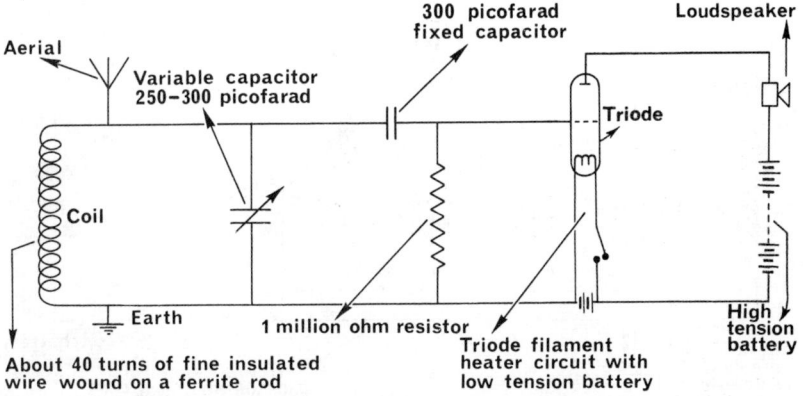

Figure 32.7. A simple radio receiver using a triode thermionic valve as both detector and amplifier. The values marked will help the reader to construct a receiver for amplitude modulated medium wave broadcasts. The potentials of the two batteries will depend on the values recommended by the makers of the particular triode selected for use in the receiver.

A cathode ray tube or oscilloscope makes use of a very narrow beam of electrons generated in an 'electron gun'. This beam is bent vertically and horizontally by two separate electric fields applied to the beam. The beam continues to the end of the tube and it forms there a glowing spot of light where it strikes the fluorescent screen. The cathode ray oscilloscope has very many uses in the study of electronics, but probably is best known as the 'tube' in a television set and as the 'screen' in radar.

Television. The main parts of the visual broadcast operation of a television transmitter and receiver are shown in Figure 32.10. The camera lens forms an image of the object to be televised on a light sensitive plate where thousands of tiny photo-electric cells free electrons on their surfaces in quantities proportional to the intensity of the light falling on them. An electron gun shoots a narrow beam of electrons back and forth across the plate scanning the whole surface by a series of 625 separate horizontal lines and this scanning is repeated 25 times every second, producing several million impulses a second due to the charges present on each little photo-electric cell at the time the scanning beam passes. These impulses are amplified and made to modulate a UHF carrier wave which is then broadcast. At the receiving station the incoming impulses are treated as they would be in a radio receiver and amplified and detected. They are then used to control the electron beam of a cathode ray tube as it scans the screen and thus reproduce there an image of the original object. Each time the scanning of the whole surface takes place, the object moves slightly so that the final image on the TV screen changes slightly. This movement of the image takes place 25 times a second so that the eye imagines it sees a continuously moving image, as it does when it looks at the changing images on a cinema screen.

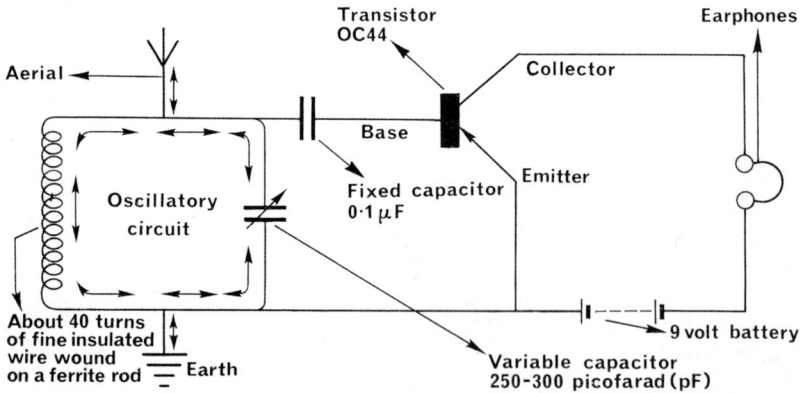

Figure 32.8. A single transistor receiver. The transistor acts both as a detector and an amplifier. The weak incoming signal from the aerial–earth circuit controls, by means of the transistor, the electric current flowing from the battery through the headphones. This arrangement will give a much louder reception in the headphones than will be heard with the circuit arranged as in Figure 32.5.

Radar means *ra*dio *d*etection *a*nd *r*anging. We know that the timing of an echo in sound from a distant cliff enables one to calculate how far away the cliff is situated. In a similar way a radio wave reflected from an object is used to determine the position of that object. Because radio waves travel very rapidly (3×10^8 m s^{-1}) this method can be used to track even fast moving aircraft. The radar transmitter sends out in a narrow beam short pulses of microwaves. The radio echo from the object returns to the radar receiver where it is amplified, detected, and passed on to a cathode ray oscilloscope. Both the original pulse and the reflected pulse are applied to the vertical set of deflecting plates of the same cathode ray oscilloscope. The horizontal deflecting plates trace out a straight line across the screen that is marked in distance measurements and not in time. Thus the gap between the two pulses or 'pips' on the screen shows the actual distance that the reflecting object is away.

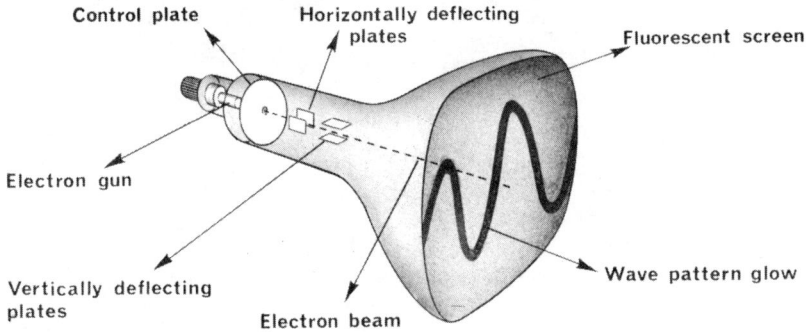

Figure 32.9. A cathode ray tube showing a visual pattern of the combined wave forms applied to both pairs of deflecting plates.

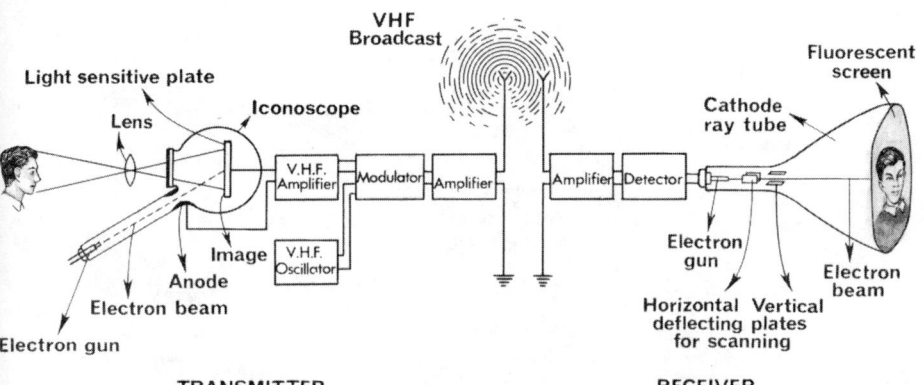

Figure 32.10. The transmission and the reception of a visual image by television.

Rotating radar transmitters and receivers are now made so that they can trace out a map or plan of the area searched. Used on ships they show a circular map which looks like a bird's-eye view of the area surrounding the ship and thus warns the ship's navigator even in darkness, in fog, or in rain that he is near another ship, rocks, or icebergs.

Radio-astronomy is a new science that has developed out of the use of radar. It has presented man with the possibility of adding much to his knowledge of the Universe. Radio microwaves as used in radar can be used to 'look' at the stars, the planets, and meteorites. A radio signal to the moon and back again

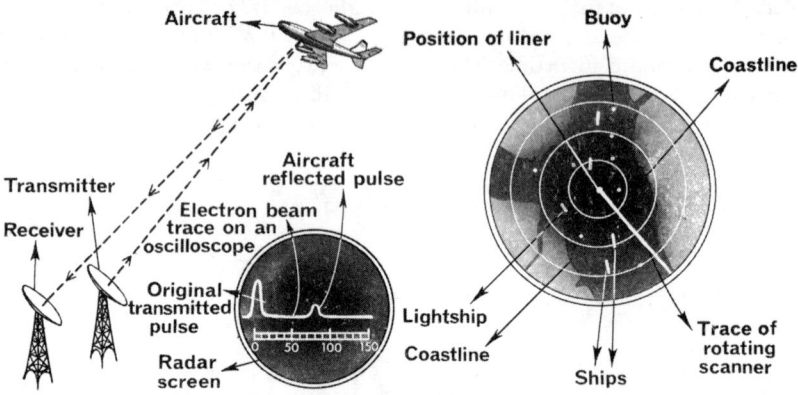

Figure 32.11. A radar set tracking the approach of an aircraft. By this means air traffic controllers know the position of an aircraft even though they cannot see it with the naked eye.

Figure 32.12. The radar screen as seen by the navigator of a liner as he proceeds along the Solent on a foggy night. The navigator knows how to interpret the spots he sees on the screen. He knows those which are ships, which are warning buoys, and which are light-ships. The coastline is just visible on either side as a faint line. The bright rings of light on the screen are produced by electrons in order to show at a glance the distance any object is away.

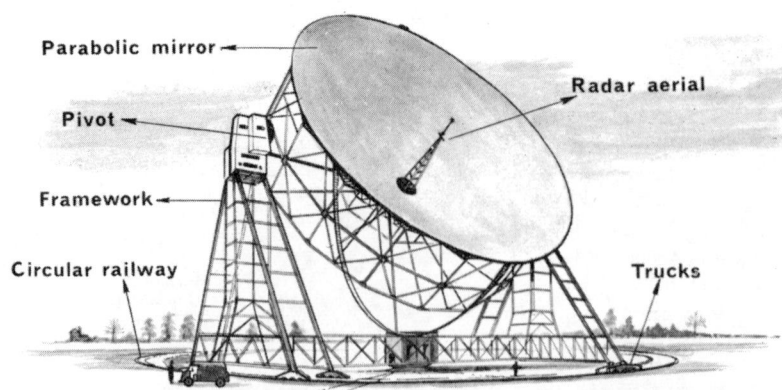

Figure 32.13. A large radio-telescope mounted on electrically driven trucks on a circular railway. The huge parabolic mirror can tilt on pivots so that it can point in any direction. The small radar aerial is mounted at the focus of this parabolic mirror. The diameter of the mirror can be 100 metres of more.

takes about 2½ seconds for the journey. The giant radio-telescopes constructed in different parts of the world reflect microwaves from a radar set in their huge wire mesh 'mirrors', and with them it is possible to track the course of man's interplanetary rockets and his artificial satellites. Strong radio waves originating in outer space have been received by radio-telescopes but so far we know little about these waves.

Recording and the reproduction of sound is accomplished today almost entirely by electrical methods involving electronics.

There are several types of *microphone* in use for converting sound waves into the corresponding varying electric currents. The quality of reproduction varies considerably between the different types.

In all cases the sound wave causes a corresponding motion of a diaphragm, which is coupled directly to the unit which generates the electric currents.

Three types of microphone for varying the electric currents are described. The *carbon granule* type, discussed already in chapter 31, is used mainly for telephone receivers. The *crystal* type consists of two slices of crystal cemented together and coated to protect them from dampness. An e.m.f. is generated across the opposite faces of the double crystal when it is deformed by the vibrations of the diaphragm. The third type is *electromagnetic*; in it a small coil is moved to and fro in the field of a permanent magnet. In one model the coil is replaced by a metal ribbon which acts as both the diaphragm and the conductor moving in the magnetic field. This is known as a *ribbon* microphone.

The same working principles are employed in the gramophone *pick-up* as in the microphone. The needle or stylus of the pick-up provides the vibrations and these are used to distort a crystal or to vary the magnetic field cutting a coil. Thus crystal (or sometimes they are called piezo-electric) and magnetic pick-ups are in normal use in modern record players.

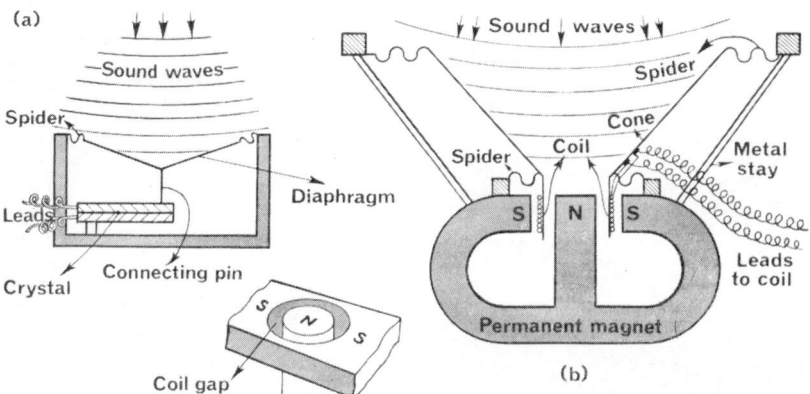

Figure 32.14. The microphone. (*a*) The crystal type—the crystal is deformed by the vibrations of the diaphragm. (*b*) The electromagnetic type—the coil moves in the field of a permanent magnet.

The loudspeaker is in principle an electromagnetic type of microphone used in reverse (see Figure 28.9). The diaphragm or cone sends out sound waves in front of the loudspeaker and these produce the sound one hears. It also sends out waves to the rear and for best results these should be absorbed inside the loudspeaker case. The greater the diameter of the cone the better the loudspeaker responds to notes of low frequency.

The *tape recorder* stores its recorded sound as variations in the way in which a finely powdered oxide of iron coated on a thin plastic tape is magnetized. During the recording operations the sound waves generate varying electric currents which then produce a variable magnetic field. This produces a magnetic pattern on the tape that moves through the magnetic field at a constant speed. 'Playing back' the tape reverses these operations. The tape is unaffected by the playback and it can be played back indefinitely without loss of quality. The tape can be cleared of its record by running it at a constant speed through the strong and steady magnetic field close to a bar magnet.

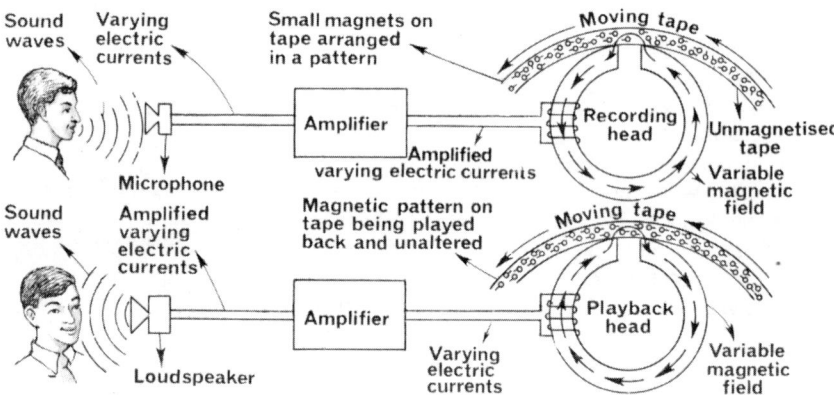

Figure 32.15. The recording (above) and the playback (below) of speech and music as magnetic patterns on tape coated with an oxide of iron.

Questions on chapter 32

1. What kind of waves are radio waves? What is the wavelength of a radio wave that has a frequency of 600 kilohertz? In what ways do radio and light waves differ? The velocity of radio waves is 3×10^8 m s^{-1}.

2. What is meant by a carrier wave and what use can be made of it?

3. What are the two methods of modulating a carrier wave? Why is one method better than the other for radio communication and especially for music?

4. What are the essential parts of a simple radio receiver?

5. Describe, with the aid of diagrams, the difference between a radio frequency and an audio frequency wave. Why cannot a loudspeaker respond to a radio frequency wave?

6. Explain how a diode can be used to 'detect' a radio wave. What additional property has a triode?

7. How is a radio receiver 'tuned' to receive a particular broadcasting station?

8. Explain, with the aid of block diagrams only how a radio receiver can pick up very weak signals and convert these into sound waves of reasonable strength in the loudspeaker.

9. What are the advantages of a transistor in radio communication?

10. Explain how a simple transistor receiver works.

11. Draw a diagram of an iconoscope and explain how the image is split up into electrical impulses.

12. How is the electron beam in a television receiver made to scan the screen of the cathode ray tube?

13. In what ways is radar similar to light as far as reflection and detection are concerned?

14. What are the practical uses of radar and how important are these to modern navigation?

15. Describe the functions of the following in reproducing recorded music and show how they are connected in the electrical circuits: (a) a magnetic 'recording head' and a 'play-back head' of a tape or wire recorder, (b) an electromagnetic pick-up of a gramophone, (c) an audio frequency amplifier, and (d) a loudspeaker.

Answers to questions

CHAPTER 1 (p. 8)
  5. 7.5 **6.** 107.1 cm **7.** 8 m
CHAPTER 2 (p. 17)
  2. 0.5 m **4.** 2.67 cm, 16.6 cm from mirror
CHAPTER 5 (p. 40)
  3. 1 cm, 15 cm from lens **4.** 6 **5.** 32 cm
  6. 4.5 cm from the scale, virtual **12.** 1.75 m
CHAPTER 6 (p. 45)
  3. 10 m **4.** Outside illumination is 25 times the illumination in the dark room **5.** Luminous intensity of motor-car lamp is 36 times that of the torch lamp **6.** 55.5% **7.** 15 m **8.** 25 m
CHAPTER 7 (p. 53)
  1. 122.5 m, 10 s **2.** Yes **3.** (a) 70 km, (b) 70 km h^{-1}
  4. No **5.** 9.6 m s^{-1}, 9.5 m s^{-1}, 8.0 m s^{-1}, 6.7 m s^{-1}, 6.2 m s^{-1}, yes
  6. 8.7 m s^{-1} **7.** 1.4 s, 13.7 m s^{-1} **8.** 2 s **9.** 16.9 m s^{-1}
CHAPTER 8 (p. 59)
  6. (a) 800 N, (b) 880 N, (c) 640 N **7.** 120 N **8.** 2.5 m s^{-2}
  9. 9.1 m s^{-2}, 1 000 N, 55 m **10.** 2.5 m s^{-2}, 8 s, 80 m
CHAPTER 9 (p. 68)
  1. 265 N, 19° **2.** 11.6 N **4.** 2.4 m **5.** 87.5 N
  6. 1.125 kg **7.** 2 m **8.** 24 000 N on the nearer pier, 6 000 N on the further pier
CHAPTER 10 (p. 72)
  1. 500 N **2.** The loaded lorry weighs 5 times the weight of the motor car
  3. (a) 3 cm, (b) 0.33 cm
CHAPTER 11 (p. 83)
  1. 0.85 g cm^{-3} **11.** 0.8 g cm^{-3}, 800 g, 800 cm^3 **12.** 750 g, 750 g, 600 cm^3
  13. 20 cm^3, 4.5 g cm^{-3} **14.** 23.5 cm^3, 53.8 g
  15. 1.6 × 10^5 cm^3, 1.5 g cm^{-3}
  16. (a) 100 cm^3, (b) 1.15 g cm^{-3} **17.** 90% **18.** 2.2
CHAPTER 12 (p. 92)
  10. 39 m^3 **11.** 3 × 10^{-2} m^2
CHAPTER 13 (p. 96)
  1. (a) 1 600 W, (b) 2 000 W, (c) 80%, (d) 2.6 **3.** Heavier
  5. 165 000 J, 1.23 **7.** 6.7 × 10^3 **8.** 16 500 J, 0.037 **9.** 8.85 × 10^4
  10. 1.6 × 10^6 N, 1.6 × 10^6 N in opposite direction

CHAPTER 14 (p. 101)
   1. 80 m, 4   **2.** 62.5%   **3.** 143 N   **4.** 12.5 N
   5. 62.5   **12.** 40 N   **13.** 500 J, 33.3%   **14.** 40 N
CHAPTER 16 (p. 116)
   2. 2×10^{-5} °C^{-1}
CHAPTER 19 (p. 143)
   4. 1 680 J °C^{-1} **5.** 140 J kg^{-1} °C^{-1} **6.** 825 J kg^{-1} °C^{-1}
   9. 2.18 J kg^{-1} **10.** 3.3×10^6 J released   **12.** 89
CHAPTER 21 (p. 158)
   2. 252 Hz
CHAPTER 22 (p. 165)
   2. 0.61 m   **10.** 330 m s^{-1}
CHAPTER 23 (p. 169)
   1. 664 m
CHAPTER 27 (p. 203)
   5. (a) 10Ω, (b) 1.6Ω   **6.** (a) 6 V, (b) 2 V   **7.** 1.5 V
   10. 12, 72 W
CHAPTER 28 (p. 211)
   10. (a) 20/99Ω, (b) 1 480Ω
CHAPTER 29 (p. 220)
   4. (a) 10 A, (b) 20Ω, (c) 30   **5.** 157.5   **6.** 5 A, 40Ω
   7. 5, 3   **8.** (c) 2 A, (d) 125Ω, 28   **9.** (a) 2 A, (b) 2.5, (c) 105
   10. (a) 7 A, (b) 10.5, (c) 61.5, (d) 16
CHAPTER 31 (p. 242)
   10. 40–1, 1.8 W, 555 h 33 min   **14.** (a) step-down, (b) 20–1
CHAPTER 32 (p. 255)
   1. 5×10^2 m

Index